HOW CHILDREN LEARN SCIENCE

HOW CHILDREN LEARN SCIENCE

conceptual development & implications for teaching

RONALD G. GOOD

The Florida State University

MACMILLAN PUBLISHING CO., INC.
NEW YORK
COLLIER MACMILLAN PUBLISHERS
LONDON

Printed in the United States of America

Macmillan Publishing Co., Inc.
866 Third Avenue, New York, New York 10022

Collier Macmillan Canada, Ltd.

Library of Congress Cataloging in Publication Data

Good, Ronald G
How children learn science.

Includes bibliographies and index.
1. Science—Study and teaching (Elementary) I. Title.
LB1585.G64 372.3'5'044 76–5899
ISBN 0–02–344640–4
Printing: 1 2 3 4 5 6 7 8 Year: 7 8 9 0 1 2 3

To butterflies,
young children,
and others
whose freedom is their life.

PREFACE

Cognitive development research and theory are gradually becoming an important area of concern for educators, particularly in science and mathematics. The main purpose of this book is to identify important science and mathematics-related concepts and to explain the nature of the development of corresponding mental structures that enable the individual to comprehend the concepts or processes in question. Implications for science teaching and curriculum development are then discussed.

The theory of cognitive development is de-emphasized in favor of descriptions of the experimental interview situations that were used to collect information on the development of logical thinking. Just as one cannot discuss relativity theory without acknowledging the overwhelming influence of Albert Einstein, cognitive development work is almost always associated with Jean Piaget. Because of this fact, much of the content of Parts One and Two is derived from Piaget's work since the 1920s. His closest collaborator, Barbel Inhelder, and others who have used Piaget's ideas as a springboard for their own work are mentioned in conjunction with specific studies on children's thinking. Part One is a brief overview of the gradual evolution of Piaget's experimental methods and areas of investigation since the early 1920s. Examples of his interview methods during the 1920s and early 1930s are contrasted with later techniques involving manipulative materials. Enough detail is provided in some of the descriptions to allow the interested reader to try an interview or two with some children.

Part Two contains the main thrust of the book. It is here that dozens of experimental interview procedures with children are presented. General results of the experiments are also described, usually in terms of the chronological ages at which many or most of the children achieve a beginning operational understanding of the concepts involved. Part Two is divided into the two main sections "Concept Development at a Concrete Level" and "Concept Development at an Abstract Level." The concrete, stage includes the ages up to eleven to twelve years; abstract thought begins to appear after this point. The "preoperational" stage in Piagetian theory is usually identified with the age range of two to six to seven years. However, this can be misleading since for many concepts such as area, weight, and volume, children do not usually become "operational" until they are well over seven years of age.

Abstract thinking or, as Piaget has termed it, formal operational thinking is characterized by the ability to go beyond concrete reality to the hypothetical. All possible combinations of events, either real or possible, are considered in an exhaustive way and empirical proofs can be offered. Such content or processes as probability, combinations, proportions, and identifying and controlling variables are functional and interrelated and problems both real and hypothetical can be conceptualized. The concept of definite "stages" of development as suggested by Piaget is de-emphasized in this book in favor of a concept of continuous cognitive growth.* The question of whether a majority of adolescents and adults are capable of logical or abstract thought is raised, and the cognitive requirements of various science concepts and principles are compared with results of related studies.

Although Part Three in this book deals with a summary of the *theory* of cognitive development, the emphasis is on a *practical* approach to understanding the findings of various studies and applying them to curriculum and instruction in science. The cognitive characteristics of children should be a very important base for decision making in the field of science education. To make rational decisions about content, approach, and the like in any field of endeavor, one must have a working knowledge of the factors affecting the outcome. Science education for children is no exception. Where "thinking" is a desired outcome, the teacher must have an understanding of the general cognitive characteristics and range of abilities of the children. Many textbooks and materials that have been developed for science instruction assume a level of thinking that is not available to many or, in some cases, to all children in a given classroom. This fact is illustrated in Part Four where examples from eight science textbook series are described in terms of conceptual levels and compared with corresponding cognitive developmental levels. Science "processes" are also described, and their conceptual levels are analyzed in terms of developmental levels.

Part Five includes a look at five different teacher roles and compares each role with four conditions that grow out of developmental theory and the nature of science. Examples of science activities are described, following the conclusions established earlier in the book, on the nature of children's thinking, teacher roles, and the "process" nature of science.

Part Six concludes this book with suggestions for helping children learn science. The suggestions are intended to be consistent with the earlier material on children's thinking and instructional strategies.

The four appendixes are referred to at appropriate places throughout the text and should be read in conjunction with certain material. However, Appendix D is a *self-test* and, if it is to serve any useful purpose, it should be taken *before* you begin to study the material in this book and then once again after you have completed your study. A comparison of

* Refer to Appendix C for paper entitled, "The Myth of Stages in Cognitive Development."

your responses should provide a useful guide to the learning that has occurred.

It is hoped that this book will enable the reader better to understand the rather complex field of cognitive development and incorporate the ideas in an effort to improve science teaching and learning. The fact that very little emphasis is placed upon the affective domain as an important determinant in science learning should not mislead the reader into discounting the importance of attitudes, values, feelings, and the like. Plenty of evidence supports the critical nature of the role of the affective domain in all learning. It is just that a comprehensive look at the cognitive domain is such a monumental task alone, that fair treatment of the role of the affective requires and deserves a separate study.

As a final note, I would like to thank the many colleagues and former students who have influenced my thoughts over recent years and thus the contents of this book. Very special thanks go to my wife, Elaine, for her dedication and understanding in assisting me in completing the manuscript.

R. G. G.

CONTENTS

part one

THE NATURE OF COGNITIVE DEVELOPMENT RESEARCH

part two

SCIENCE-RELATED CONCEPTS AND THEIR DEVELOPMENT

part four

SCIENCE CONCEPTS AND PROCESSES: COGNITIVE REQUIREMENTS

part five

THE ROLE OF THE TEACHER

appendix c

appendix d

part one

THE NATURE OF COGNITIVE DEVELOPMENT RESEARCH

objectives

It is intended that this section will help the reader to

1. Understand the nature of the clinical interview as a method of finding out about a person's conception of causality.
2. Interpret different types of responses to questions about causality.
3. Understand the nature of interviews with manipulative aids.
4. Identify specific interview procedures associated with such concepts as number, class, space, and the like.

The material provides a general introduction for the more extensive discussion of intellectual characteristics of children and adolescents in Part Two.

introduction

Any adult who has taken some time out from the seemingly "inherited" task of coaching children to "behave right" and, instead, observed the natural behaviors of children engaged in self-structured activities has engaged in a form of cognitive development research. Patient and careful observations of children in either "natural" or contrived environments characterize the essence of much of the research on the development of children's thinking that has been done during the twentieth century. One individual in particular, Jean Piaget, has become the leader in the field of cognitive development research by virtue of his many books, papers, and the like that have appeared since the early 1920s. His efforts, which span six decades, have provided the impetus for many other researchers to discover more about how the human being acquires thought mechanisms required for logical thinking.

As in the field of science, "searching for patterns" characterizes the work of individuals engaged in cognitive development research. The seemingly random thought processes expressed in the behavior of children begin to become more predictable through careful observation of and interaction with children of various ages. The mechanisms of the growth of scientific-type thought including such concepts as space, time, number, weight, motion, and many more can become more visible as systematic analyses are made of the child's interaction with his environment.

The purpose of Part One is to identify specific examples of data collection in cognitive development research and then look for patterns that will allow for a general classification of various levels of thinking. For the reader who is already somewhat familiar with experimental methods used in cognitive development research and the characteristic stages that have been formulated from the data, Part One can be treated briefly in favor of later material that describes specific concepts and principles in more detail.

professional background of Jean Piaget

An understanding of the person can often lead to a more accurate understanding of that individual's ideas. In the case of Jean Piaget, we can see in his early interests in biological studies a guide to the intent of his later work in what has been termed *genetic epistemology*, or a study of the mechanisms of the growth of scientific-type knowledge.

The town of Neuchâtel, Switzerland, was the birthplace of Jean Piaget on August 9, 1896. Piaget's keen interest in biological studies as a youngster is emphasized by a paper that he published in the *Journal of Natural History of Neuchâtel* when he was only ten years old. The object of his study reported in the paper was an albino sparrow. It was apparently his intense perception of the various behaviors of the sparrow and the resultant thoroughness that he exhibited in his paper that caused the publishers of the journal to view young Piaget's first formal report as worthy of publication. Interestingly enough, the same perceptiveness in observing and interviewing children was later to become Piaget's trademark in the field of cognitive development research. Piaget's interest in biology continued into his teenage years and he continued to publish papers on his studies of mollusks and related zoological topics.

A vacation with his godfather, who was a scholar in his own right, started a broadening of Piaget's interests into philosophy and logic, although biological studies continued to dominate his scholarly pursuits. So we can see a kind of dual interest developing: biological development of life and philosophical nature of knowledge. This combination of interests prepared Piaget in a rather unique way to begin his studies, a few years later, in which he delved into genetic or biological explanations of the growth of a person's ability to acquire scientific-type knowledge, or, in short, to think logically.

At the age of twenty-one, Piaget completed the work required for a

Doctor of Philosophy degree in the field of zoology at the University of Neuchâtel. The next couple of years were not very productive in terms of identifying problems worth pursuing, until in 1920, when he was offered a position at Alfred Binet's laboratory in Paris. Piaget was asked to develop a standardized French version of Burt's reasoning tests in order to be sure that differences in performance were related to a subject's ability rather than the manner in which the questions were presented. It soon became apparent to Piaget, however, that the rigid structure of standardized tests did not allow the interviewer to pursue important clues to the thinking abilities of children. The methods that various children used in responding to questions became the central focus rather than merely how many questions were answered correctly. The less structured clinical approach allowed Piaget to follow the child's line of thinking in any direction.

During this time, Piaget attended lectures by Freud and Jung. Freud's influence caused Piaget to begin a study of children who were labeled "abnormal," in the hope that he would gain more insight into normal cognitive processes. It was during this period that Piaget developed an interview procedure that was to become the basis for most of later research in cognitive development. Since the "abnormal" children were deficient in verbal abilities, Piaget began using manipulative materials in the questioning procedure to assure a more accurate means of communication.

Piaget's concurrent reading in logic caused him to notice that a great many children were unable to carry out even very elementary logical thinking operations. These first experiments with children's logical thinking abilities would eventually lead Piaget to make extensive use of the language of logic to communicate his findings (much to the chagrin of initiates to his work).

the nature of Piaget's experimental methods

In 1921, he was offered the job of Director of Studies at the Rousseau Institute in Geneva, which he accepted. During the next few years he engaged in research on children's thinking abilities that soon caused him to become recognized internationally.

Studies of language development and interpretation of physical causality, moral judgment, and reasoning ability were carried out and later described by Piaget in five different books (1923, 1924, 1926, 1927,

1932) and in various journals. His interviewing techniques relied very little on manipulative aids as he had done a few years earlier with "abnormal" children, but he gradually realized the drawbacks of a purely verbal type of interview. The logic that children applied to situations involving concrete objects oftentimes seemed different from their logic as applied to purely verbal statements.

It was also during this period in the late 1920s that Piaget's well-known studies of intellectual development in infants were carried on with his own children. Two daughters and a son became the subjects of intensive study by both Piaget and his wife. From birth until the age of two, the Piaget children were observed in an attempt to piece together behaviors that would show a pattern of cognitive development during the first two years of a child's life, later to become known as the "sensory motor" stage of development. This study by Piaget of development during infancy clearly pointed out the primary importance of a child's actions rather than his or others' language in his thinking (1936 a and b). After these experiences, Piaget realized the critical importance of using concrete objects during interviews with young children, and his later work reflects a modification in interviewing techniques.

During the 1930s, Barbel Inhelder and Alina Szeminska worked closely with Piaget in studies dealing with number and quantity concepts. The research was reported in various articles and two books (1941a and 1941b). Significantly, this work with number and quantity led to further research that continued to focus on specific science-type concepts. Also during the early 1930s, Albert Einstein was to provide the impetus for investigations of the child's understanding of time, movement, and speed (1946 a and b). Out of experiments on perception, two books were published in 1948 that reported the child's understanding of space and geometry (1948 a and b).

Two of Piaget's many books are of particular interest to our understanding of children's development of science-type concepts covering the range of the elementary and middle school years (1955, 1959). Perhaps more than his other works mentioned so far, these two books reflect the desire on Piaget's part to communicate his findings by way of the language of logic. The formal language of logic is not susceptible to the variety of interpretations that is characteristic of our conversational language. It is interesting to note that the 1955 book represents the only major research effort of Piaget and his co-workers at the J. J. Rousseau Institute that is concerned with adolescent thought.

The interests and understanding of all of us change as we grow more knowledgeable about our work, and Piaget's books and articles published since the early 1920s reflect such changes. After studying the nature of children's thinking for over thirty years, Piaget gradually returned to the study of genetic epistemology. More recent books by him (1967) and jointly with Inhelder (1968) reflect his trend away from studying specific scientific concepts toward the process of constructing a theoretical framework to explain epistemological problems in terms of biological factors.

example interviews and results

Interviews without Manipulative Aids

Interviews that rely exclusively on the conversational mode between the child and the interviewer have some advantages but many drawbacks. Our conversational language can have many different interpretations. In spite of the problems inherent in collecting data using only the flowing conversational technique, many valuable insights can be gained. Two very interesting early books by Piaget, *The Child's Conception of the World* (1926) and *The Child's Conception of Physical Causality* (1927), contain insightful conversations with children of ages five to fifteen years. In the 1926 book, questions dealing with how people think, the origin of names, dreams, the consciousness of things, concepts of life, and other "worldly" conceptions are used to determine "reality" as seen by children of various ages. Piaget attempted to find the extent to which children distinguish between the external world and an internal or subjective world. The later book, dealing with physical causality, investigates the child's conceptions of air, clouds, sun and stars, water currents, force, floating, shadows, and other common objects and events in our environment. A few examples of more recent interviews (1970s) should help to communicate how such conversations can be used to gain insight into children's thinking.

Interview with Tom

INTERVIEWER: Where does the wind come from?
TOM: The East. South.
INTERVIEWER: Where is it now?
TOM: All around us. Can't you feel it?
INTERVIEWER: What is it made of?
TOM: Air. Like, we have air. We can blow and make the wind too.
INTERVIEWER: Where did the wind come from in the first place?

TOM: Well, you see, God gave men air so they could breathe, and he wanted to help them out a little more. Like, you know sailboats? Well, the air moves them. The air pushes, and God helped us out so we could ride sailboats.

INTERVIEWER: Why does it move?

TOM: God moves it.

INTERVIEWER: Does it feel us when it blows against us?

TOM: Yes, it touches us.

INTERVIEWER: Is it alive?

TOM: Yup.

INTERVIEWER: Where does it go at night?

TOM: To the . . . West . . . North.

INTERVIEWER: Is it always in the same place?

TOM: No, it moves around.

[*The interview now changed to a discussion of the sun and moon.*]

INTERVIEWER: Where is the sun?

TOM: Up there [*pointing*] behind those clouds.

INTERVIEWER: Is it far away?

TOM: Yeh, it's *way* far away.

INTERVIEWER: What does it do?

TOM: It shines. It gives us light so we can see.

INTERVIEWER: Where did it come from?

TOM: God made it for us.

INTERVIEWER: What is it made of?

TOM: Fire. And the Devil is too. You see, when God made the sun he made the Devil too.

INTERVIEWER: Are they in the same place?

TOM: No! See, the sun is up there and the Devil is down there!

INTERVIEWER: Where does the sun go at night?

TOM: It goes to sleep. And then the moon wakes up and *it* shines!

INTERVIEWER: Where does the sun go to sleep?

TOM: Well, you see those clouds up there? They go all around and cover him up like this [*forms circle with arms*], and he's all covered up and he falls asleep.

INTERVIEWER: How does the moon wake up?

TOM: Well, God made a man, and his name was 'Econ' . . . kind of a pretty big name, but that's what his name was, all right! And, Econ, he's a *very* fast runner, and when it's time for the moon to wake up he runs as fast as he can and wakes him up . . . and then it's time for the sun to wake up, so he goes back and forth, back and forth. . . . Like this! [*demonstration*]

INTERVIEWER: What is the moon made of?

TOM: Green cheese. [*Looks at me with a kidding smirk.*] No, that's not really what it's made of. You know what? It's made of *laser* beams. Yup. . . . They're scientists' greatest invention of all times. They're the greatest defense weapon we ever had. And they helped the astronauts to land on the moon. God invented the sun and the moon, and I know who invented the light.

INTERVIEWER: Who?

TOM: Edison.*

Tom was a six-year-old in first grade at the time of this interview and was considered by his teacher to be "very aware" of his environment. It is not difficult to see that reality as made up by Tom's mind is different from "adult-type" reality. For questions dealing with the origin of things, Tom consistently referred to God as the originator. In responding to questions about the wind, sun, and moon, Tom also used God in explanations about how something occurred. God makes air move so man can use his sailboats. God makes the sun shine so we can see. In a sense, God is seen as the servant of man, to do that which we need. Air exists so we can breathe; light exists so we can see. Tom's thinking is very egocentric throughout the interview, reflecting an inability to make up explanations not involving his own desires. When a child attributes the creation of the sun to men rather than a God, the child is still reflecting egocentrism. The sun is still "made for us."

Piaget (1926) identified three characteristics of young children's thinking that tend to dominate explanations of objects and events in their environment. *Animism* was the term applied to the tendency of attributing a consciousness to things such as the sun, moon, trees, stones, air, and the like. These "things" could act voluntarily and thus move, change color, and so on. *Artificialism* was used by Piaget to denote the child's tendency to believe that things are made for man's benefit. In Tom's case, air was made so we could breathe. Things exist or occur because man needs them. Finally, *magic* was used to characterize explanations where man caused things to happen, that is, movement of clouds, movement of sun, and the like. Until the ages of about seven to eight years,

* Thanks to Dr. Gwendy Camp, former graduate student in science education at Florida State University, for this interview.

children tend to explain causality in their environment in terms of combinations of magic, animism, and artificialism. After seven to eight years, there is a gradual progression toward more "objective" explanations of causality.

Another interview at this point might help to identify other examples of how one can pursue causal thinking in young children. This interview with Teena, a six-year-old, took place at the edge of a wooded area in the early evening with the sun "hidden" from view behind the trees.

Interview with Teena

INTERVIEWER: I want to ask you some questions about what you think about things and you just tell me what you think, okay?

TEENA: Okay.

INTERVIEWER: When you look out at the trees, what kinds of things do you notice about the trees.

TEENA: Leaves are growing on them.

INTERVIEWER: Anything else?

TEENA: Flowers; tulips are growing on them.

INTERVIEWER: Okay, just tell me all the things you notice about the trees.

TEENA: New limbs are starting to grow.

INTERVIEWER: Anything else?

TEENA: No.

INTERVIEWER: Okay. Do you notice any big trees and any little trees?

TEENA: Little trees are beginning to grow.

INTERVIEWER: Point to a tree that you say is a little tree. Do you see a tree out there that is a little tree?

TEENA: That one over there [*pointing to a tree that is about 4 feet tall*].

INTERVIEWER: Okay. Do you see one which is a big one?

TEENA: That one right there [*pointing to a tree about 35 feet tall*].

INTERVIEWER: Okay. What do you think causes some trees to be big and some trees to be little?

TEENA: The more they get water the more they grow.

INTERVIEWER: Do you know of anything else that causes the big ones to be big and the little ones to be little?

TEENA: They're in different sizes.

INTERVIEWER: Okay. Anything else you can think of that causes the big ones to be big?

TEENA: Nope!

INTERVIEWER: Okay. Just the water; if they get more water, they get bigger and—what was the second thing?

TEENA: The more they get, the bigger they get and the less water they get, the little they get.

INTERVIEWER: I see. Well, suppose you wanted to find out how to make a tree grow bigger and bigger? What would you do to find out?

TEENA: Water it.

INTERVIEWER: You'd water it.

TEENA: . . . and give it fertilizer.

INTERVIEWER: You'd give it fertilizer. Do you think fertilizer would make it grow bigger?

TEENA: Yes, when you plant it.

INTERVIEWER: Okay. If you see that tree that you pointed to—that big tree and that little tree, why do you think that one is bigger than that one [*pointing to the two trees the child has previously pointed out*]?

TEENA: Because it had more water and had a bigger seed.

INTERVIEWER: Do you think if it would have a bigger seed, it makes the tree grow bigger?

TEENA: Yes.

INTERVIEWER: If you had two seeds and one was little and one was big. . . .

TEENA: One would be little and one would be big.

INTERVIEWER: How big do you think the seed was for that tree right there [*pointing to the 4-foot tree*]?

TEENA: A weenie, weenie one.

INTERVIEWER: Okay. Well, suppose you had some seeds to plant, how would you find which one is going to make a big tree and which one is going to make a little tree?

TEENA: How big they are.

INTERVIEWER: Okay. Have you ever seen the trees when they're moving around?

TEENA: Yes, when you take them out of the ground and you plant them some place else.

INTERVIEWER: They move around then. Have you ever seen them when the limbs are moving around and nobody is moving them?

TEENA: Yes. The wind is pushing them.

INTERVIEWER: You think the wind is pushing them when that happens. What do you think makes the wind push them?

TEENA: Uh—the trees are blowing.

INTERVIEWER: The trees are blowing, so, let me see if I have it right now—what you said. When the limbs are moving around and nobody is moving them, the wind is pushing them?

TEENA: Yes.

INTERVIEWER: And what was it you said is causing the wind to move?

TEENA: Well—uh, big trees; limbs are moving because the tektor of it makes the trees move and the wind starts windy.

INTERVIEWER: What was it that makes the wind move?

TEENA: The tektor from the tree. The tektor makes the tree move back and forth.

INTERVIEWER: What is the tektor? What is that?

TEENA: Uh—I don't know but I just—it's another word, I think.

INTERVIEWER: Am I saying it the way you did?

TEENA: I don't know how to take tektor—people breathe . . . I don't know how to say that word, so that's why.

INTERVIEWER: You said people breathe? Is that what you said? Does that have something to do with tektor?

TEENA: Yes.

INTERVIEWER: So what does it do when people breathe? Does that have something to do with what we were talking about?

TEENA: If there wasn't any air people would be dead.

INTERVIEWER: If there were no air people would be dead.

TEENA: They couldn't breathe and their heart would get hot and they would die.

INTERVIEWER: If they couldn't breathe their hearts would get hot and they would die? So what does the air do when you breathe it?

TEENA: It helps your heart beat.

INTERVIEWER: It helps your heart beat? How does it help your heart beat?

TEENA: Because when you open your mouth it goes in, and works around in your heart.

INTERVIEWER: It does what in your heart?

TEENA: Works around in your heart.

INTERVIEWER: Works around in your heart? Do you know of anything that looks like your heart?

TEENA: I mean in your brain.
INTERVIEWER: Oh, in your brain?
TEENA: Yes.
INTERVIEWER: The air works around in your brain?
TEENA: Yes.
INTERVIEWER: And it makes your heart beat?
TEENA: Yes.
INTERVIEWER: So if you didn't have any air, what would happen?
TEENA: You would die.
INTERVIEWER: You would die?
TEENA: And your heart would get hot.
INTERVIEWER: It would get hot?
TEENA: Yes.
INTERVIEWER: What is it that would make you die?
TEENA: "Tektor" from the tree . . . if you didn't have any trees you wouldn't have any tektor and it would stop your breathing.
INTERVIEWER: So if you didn't have any trees you wouldn't have any tektor and it would stop your breathing?
TEENA: Yes.
INTERVIEWER: Okay. Have you ever heard of oxygen?
TEENA: Yes. That's what I was trying to say.
INTERVIEWER: Okay. So when you said tektor you meant oxygen? Is that right?
TEENA: Yes. Tektor is another word for oxygen.
INTERVIEWER: Oh, I see. How did you find out about tektor and oxygen and things like that?
TEENA: Cause they sound alike because and are . . . [*not understood*].
INTERVIEWER: They sound alike and are what?
TEENA: In their voices.
INTERVIEWER: They sound alike in their voices?
TEENA: Yes.
INTERVIEWER: Okay. Have you ever thought about the clouds? You've seen the clouds, haven't you?
TEENA: At night-time when the sun goes down they turn colors and they go away and the sun makes pink and yellow and white and blue.
INTERVIEWER: Yes. So you've looked at them when the sun's going down. Well, what do you suppose makes them turn pink, yellow, and all that?
TEENA: The sun when it goes down. It makes a little light

spurting out of the sun and the sun makes different colors because it's nighttime and the night makes different colors when a little bit of sun comes out.

INTERVIEWER: Hum, that's interesting. Just tell me all about that, all that you can think of.

TEENA: That's all I know.

INTERVIEWER: How did you happen to know that? Did you think of that or did someone tell you that or what?

TEENA: I just thought of that.

INTERVIEWER: You just thought of it?

TEENA: Yes.

INTERVIEWER: Did you think of it just now or before now?

TEENA: I thought of it now.

INTERVIEWER: Just now? What about the clouds in the daytime? What do they look like when it is daylight?

TEENA: White.

INTERVIEWER: They look white? Why do you say they look white?

TEENA: Because the sun's out.

INTERVIEWER: Because the sun's out? How do you think the sun makes them look white?

TEENA: The sun looks like it's a hot place and it is a hot place except the sun's really a ball full of fire.

INTERVIEWER: A ball full of fire. Hmm. Why do you think it is a ball full of fire?

TEENA: Because it's hot.

INTERVIEWER: Does it look like a ball full of fire to you?

TEENA: We saw a movie about it at school; how it looked when you got real close to it and how it looked when you got real far from it.

INTERVIEWER: Tell me about the movie. What else was in the movie?

TEENA: Yeah, about when you look in a telescope you see some people walking on the moon that don't live on the Earth.

INTERVIEWER: Some people walking on the moon that don't live on Earth. Where do they live?

TEENA: They live on the moon.

INTERVIEWER: They live on the moon?

TEENA: Yep.

INTERVIEWER: Do they live on the sun too or just on the moon?

TEENA: If they lived on the sun they would be dead because the sun is a ball of fire.

INTERVIEWER: So you don't think they live on the sun?

TEENA: No.
INTERVIEWER: Well, what about the moon—what is it like?
TEENA: Um, it's round and it's dark up there.
INTERVIEWER: Dark up there and round?
TEENA: Yes.
INTERVIEWER: What do you suppose the moon is right now? . . .
TEENA: And they have to take their own tektor.
INTERVIEWER: They have to take their own tektor? That means the same thing as oxygen, right?
TEENA: Yes. They have to take their own oxygen.
INTERVIEWER: To the moon or to the sun?
TEENA: To the moon, because if they go to the sun they would get dead because the sun's a ball of fire.
INTERVIEWER: Oh, I see. Where do you think the sun is right now?
TEENA: Um, behind the clouds.
INTERVIEWER: Behind the clouds? What about the moon, where do you think it is?
TEENA: It's getting ready to jump out.
INTERVIEWER: Jump out of where?
TEENA: The clouds.
INTERVIEWER: You think its behind the clouds too?
TEENA: Yes.
INTERVIEWER: Do you see the clouds that the sun is behind and the moon is behind?
TEENA: No, I don't see it but I know it.
INTERVIEWER: Where do you think they might be?
TEENA: Behind the mountains or somewhere like that.
INTERVIEWER: Behind the mountains? Do you see any mountains that they're behind?
TEENA: No.
INTERVIEWER: No?
TEENA: That's all I know.
INTERVIEWER: That's all you know about that?
TEENA: Yes.
INTERVIEWER: Do you want to talk any more about that?
TEENA: No.*

Teena does not use God in any of her explanations about causality even though she had opportunities to do so. This could be explained, in part at least, by the fact that the interviewer was trying to find out

* Thanks are given to my good friend and colleague, Dr. Charles Matthews, Professor of Science Education, Florida State University, for permission to use this interview.

whether Teena could identify variables that affect the growth of plants. No questions about the origin of things were asked and this is typically the area that induces references to God as the creator of things. Teena told the interviewer that water, fertilizer, and seed size would affect plant growth. Small trees grew from small seeds and large trees grew from large seeds. She did not give any indication that she could separate the variables in terms of how each might influence the growth of trees. A simple direct relationship of small seed to small tree and large seed to large tree was used in the explanation. Examples of animism and artificialism were evident in Teena's references to the moon's "getting ready to jump out" and to air being used by people to keep from dying. She also made up explanations in direct opposition to visual reality at the moment. The moon was said to be behind the clouds even though there were no clouds in sight. When the interviewer asked Teena about the clouds, she said they were behind the mountains, in spite of the fact that there were no mountains.

The interviews with Tom and Teena represent somewhat different approaches to talking with children in an attempt to determine characteristic thinking patterns. Tom was asked repeatedly about the origin and substance of things. Teena was asked for reasons causing certain differences in plants, which was actually a test for identifying and controlling variables. A third type of interviewing without the use of manipulative aids is shown in the interviews that follow. These are primarily directed at children's interpretations of scientific-type terms or events of which they had some awareness. Eight questions are asked of children with their brief responses. No attempt was made to extend an interview by pursuing initial comments as was done with both Tom and Teena.

1. *What Is Energy?*

KEITH (8 YEARS): Energy is what makes you strong. You have it when you are healthy.

CECI (6 YEARS): It means you're strong; big people have energy, but babies don't. Plants, no . . . well, yes they do.

KENNETH (6 YEARS): Energy is good to eat like vegetables like squash and stuff. People have energy and it makes them strong.

SANDY (8 YEARS): You use energy if you're healthy and strong. You can get more if you exercise.

DARRELL (8 YEARS): It's like air and water. There aren't any other kinds.

KIM (8 YEARS): Something you get when you move around. It comes from food.

CYNTHIA (8 YEARS): It is when our body is strong and healthy and we can do things we want to do.

ALEX (9 YEARS): Some people are stronger than others, that's energy. People who play baseball are stronger than people who sit inside and read. Sun is light energy, but not electrical like a light bulb. Light doesn't have energy, it has power. Energy has atoms and little particles.

2. What Happens to Living Things When They Die?

KEITH (8 YEARS): Some get buried or eaten by termites.

PAT (8 YEARS): They go down, disintegrate.

CECI (6 YEARS): When things die, they don't come back any more.

KENNETH (6 YEARS): Don't know. Go to heaven . . . , but not all things. People do but some don't . . . some go down there. I guess they have an insect heaven.

SANDY (8 YEARS): When things die they stink so you bury them and they turn to dust.

DARRELL (8 YEARS): When things die they wilt or spiders eat them.

KIM (8 YEARS): They might get eaten. They get buried too. Sometimes animals have a special cave that they can go to. I saw a movie and it had elephants that went to a special place where all elephants die.

CYNTHIA (8 YEARS): Sometimes they lie on the ground and sometimes they get buried under the ground.

ALEX (9 YEARS): They disintegrate slowly and get eaten by ants and other animals.

3. What Happens to Wood or Paper When It Burns?

KEITH (8 YEARS): It flakes; gets made into charcoal and chalk.

PAT (8 YEARS): It goes into ashes.

KENNETH (6 YEARS): Wood goes to ashes. The rest of it goes up in smoke.

SANDY (8 YEARS): If you burn it it turns black and flies up in the air.

DARRELL (8 YEARS): Paper burns by going up into a little flame, then turning into black chalk.
CYNTHIA (8 YEARS): It turns into the ground and sinks.

4. What Is an Atom?

KEITH (8 YEARS): Never heard of atoms.
DARRELL (8 YEARS): An atom is something that is real strong.
CYNTHIA (8 YEARS): I don't know.

5. What Are Germs?

KEITH (8 YEARS): Germs are things that get on water and cups. They make people sick.
PAT (8 YEARS): Germs get on dogs. They can hurt you too.
CHARLES (9 YEARS): Germs are where you sneeze or pick up something dirty.
BRETT (9 YEARS): Germs are when dogs kiss you or when you eat after others.
KIM (8 YEARS): Tiny little bugs that we can't see. They're in dirt and stuff.
SUE (8 YEARS): Little specks that get on things when they are dirty. They make you sick sometimes.
ALEX (9 YEARS): They are inside. They live in their own private worlds. They cause disease and sickness.
JEFF (8 YEARS): Some germs are good. Some are found on snakes. They aren't found on blackberries.

6. Why Is It Hot in the Summer and Cool in the Winter?

KEITH (8 YEARS): The earth goes around the moon and they go around the sun. It's cooler when its close to the moon.
PAT (8 YEARS): God made it that way.
CHARLES (9 YEARS): It's cooler in the winter because the wind blows more.
BRETT (9 YEARS): It's hot in the summer and cold in the winter because those are just names.
KIM (8 YEARS): Because it's not as sunny.
JOHN (8 YEARS): The wind in the winter makes it cooler.
CYNTHIA (8 YEARS): The snow makes the wind cool and the sun makes flowers in the summer.

LISA (8 YEARS): It just happens that way.
DANNY (9 YEARS): The earth turns on its access. In Asia it is winter. Winter is on the other side.

7. Why Do We See the Sun Only During the Day?

KEITH (8 YEARS): The world turns and stops for a minute.
PAT (8 YEARS): God made it that way.
KENNETH (6 YEARS): Because there are too many clouds at night.
SANDY (8 YEARS): You only see the sun in the daytime because the moon doesn't shine then. The moon is over near Japan and at night they switch.
BRETT (9 YEARS): You can only see the sun in the daytime because the moon gets in the way at night.
TOM (8 YEARS): The moon has to come out at night and the sun and moon can't be out at the same time.
BECKY (8 YEARS): It hides behind the trees at night.
CYNTHIA (8 YEARS): In the day the sun is the only light. In the night the sun goes away and the moon shines light.

8. Where Does the Wind Come From?

KEITH (8 YEARS): The wind is from clouds. Trucks, cars, and planes make it. It never stops.
CECI (6 YEARS): The wind comes from trees.
KENNETH (6 YEARS): The wind comes from energy—no, I think it comes from the trees.
BRETT (9 YEARS): It comes from heaven, the sea, and the ocean. It goes behind the moon and the clouds.
DARRELL (8 YEARS): The wind comes from the sky. The oxygen makes pressure and hits it and makes it go.
BECKY (8 YEARS): Wind comes from the sky, I think. After it blows, it goes back up to the sky.
LISA (8 YEARS): Wind comes from clouds. After it blows, it disappears in the air.

No attempt will be made at this point to interpret individual responses to any of the eight questions. Some responses are examples of attempts to recall information obtained previously, whereas other responses are obviously the child's own invention. The examples are rich with animism, artificialism, and magical interpretations. Later in the book, attempts are made to relate science "content" as proposed by school textbooks, curriculum guides, and the like to research on the cognitive developmental levels of children. Some of the difficulties of interpreting children's responses, however, are treated briefly here before turning to examples of interviews with manipulative aids.

Difficulties in Interpreting Responses

Reading a child's responses that have been recorded during an interview does not allow for accurate determination of seriousness, confidence, and other characteristics of the child. The pauses, smiles, puzzled looks, nods of the head, and other nonverbal cues provide input that the interviewer must consider. Piaget (1926) reported five types of reactions by children in interview situations: (1) *the answer at random,* (2) *romancing,* (3) *suggested conviction,* (4) *liberated conviction, and* (5) *spontaneous conviction.* Though other categories might be added, this five-category system is quite useful in interpreting children's responses during interview sessions.

The *answer at random* might be used by a child who is uninterested or is tired of a particular line of questioning. An example of this is seen in the interview with Teena. Near the end of the relatively lengthy interview, Teena responded to a question about the position of the sun and moon by saying they were "behind the mountains or somewhere like that." This response was the first thing that "popped" into Teena's head and was not based on any factual observations. Subsequently she indicated that she did not want to continue the interview, which gives further assurance that her answer was merely a random reply.

Sometimes, a child will answer a question with "tongue in cheek" and not really have any conviction regarding the statement. The intent may be to fool or play a game with the interviewer or merely to hear what it sounds like to say something. In either case, the child is *romancing,* that is, responding without conviction.

Suggested conviction can be avoided by the experienced interviewer by carefully analyzing the nature of questions directed to the child. Children quite often try to find cues in adult behavior that can give direction to "preferred" responses. Verbal as well as nonverbal cues must be avoided by the interviewer so as to be sure not to influence the child's responses. To investigate a child's thinking about the origin of the sun, the question could be phrased "Who do you think made the sun?" This wording, however, is much more suggestive than "How do you think the sun began?" Piaget found that suggested responses by children could be minimized if the interviewer would first spend a few days listening to conversations among children.

A fourth category of responses was identified by Piaget as *liberated conviction.* This was evidenced by "reflection" on the part of the child after a question was presented by the interviewer, without suggestion. This reasoning process by the child precedes the response and reflects a sincere, careful attempt by the child to give a "sensible" answer.

The fifth type of response was termed *spontaneous conviction* and reflected an answer with confidence, not needing careful formulation. An example of such spontaneous conviction was given by Piaget (1926) relating interviews with boys about the mechanism of bicycles. Those children at an average age of eight years who had apparently taken careful notice of the interrelationship of various parts of the bicycle would immediately give an explanation of the mechanism. The spontaneity and conviction of the responses indicated that the children had given some thought to the problem before and had worked out a solution that seemed sensible, at least to each individual.

Interviews with Manipulative Aids

The inadequacies inherent in a purely verbal interview gradually became clearer to Piaget and his co-workers at Geneva as evidenced in their books and papers published in the latter 1930s and afterward. Two of Piaget's early books (1946 a and b), reflecting the use of manipulative aids during interviews, report work related to a number of questions suggested by Albert Einstein about 15 years before the appearance of the books. Being a physicist and the originator of the well-known relativity theory, Einstein quite naturally was interested in the development of concepts related to time, motion, and velocity. Such concepts are important throughout the field of physics and are studied in some detail in a later section along with other "science-type" concepts. This section deals with examples of various types of interviews using manipulative materials. The problem of interpreting children's responses as well as the problem associated with the child's interpretation of the interviewer's questions are both minimized considerably through the use of manipulative aids. Differences in meanings of terms, as interpreted by either the interviewer or the interviewee, is always a confounding variable in data collection in the field of cognitive development, but attempts can be made to reduce the effects. Clarification of meaning can be much easier when objects are used as a part of the process.

Interviews Relating to Time

In his studies of the child's conception of time, Piaget used a number of interesting experiments to probe the child's mind. Figure 1–1 shows the simple apparatus used to study children's conceptualizations of the sequence and duration of events.

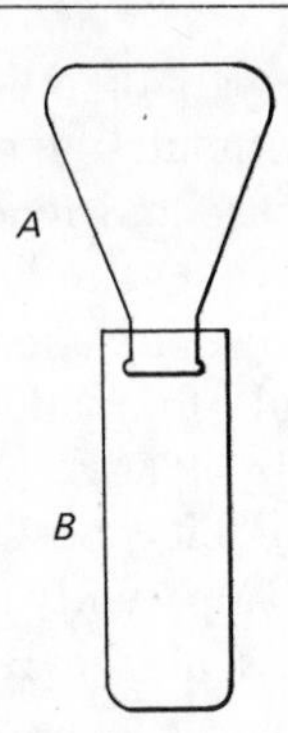

Figure 1–1. Apparatus for sequence and duration interviews.

The upper container (*A*) was equal in volume to the lower container (*B*). At regular intervals a certain amount of water was allowed to run from container *A* into container *B*. At each interval, the child was supplied with a sketch of the apparatus, without any water levels indicated, and asked to "draw in" the water level. After six to eight of these intervals, all of the liquid was contained in container *B* and the child completed the final sketch. The papers were shuffled and the child was asked to order the drawings from first to last. If the arrangement was incorrect, the child was asked about the mistakes. Each sketch containing the child's lines was then cut so as to separate container *A* from container *B*. The child was then asked to put each set in order. A number of questions were then asked of the children such as: (1) When the water level was here in container *A* (pointing to one of the sketches), where was it in container *B*?; (2) Which drawing did you make at the same time as this (pointing to one sketch of either container *A* or *B*)?; (3) Was the water here (pointing) before or after here? These and other such questions deal with the sequence of events, in this case the changing water levels in the containers. Other questions were then asked to determine the child's thinking about duration: (1) Does the liquid take as long to go from here to here (that is, from point 2 to point 3 in container *A* as compared to point 2 to point 3 in container *B*)?; (2) Does the liquid take as long to go from point 1 to point 2 in container *A*? Questions concerning the equality or inequality of time intervals are always related to a physical event when dealing with the duration of something.

Three stages were identified by Piaget as being characteristic of both sequence and duration. During a first stage, children are unable to order even the uncut sketches or to equate duration from one container to the other. During a second stage, they can order the uncut sketches but not after they are cut. Also, the children discover the inverse relationship between velocity and time, but are still unable to coordinate time intervals. Finally, the children are able to establish a serial correspondence as well, correlating durations with the correct order of events (after eight to nine years of age).

Two other concepts identified by Piaget as being important to an understanding of time are succession and simultaneity. To test for these concepts, mechanical toys were used, either having similar or different velocities. Starting times, velocities, distance traveled, and stopping times were varied, according to which concept was being studied. Questions were asked regarding which objects stopped first or last; if they moved for the same length of time; which object ran longer; which traveled the greater distance, and so on.

In other attempts to determine the development of the concept of time, measurement devices such as sand timers and watches were used, ages were compared, and tapping or rhythms were used. From all of the various experiments with sequence, duration, succession, simultaneity, and the like, it was concluded that time is the coordination of motions at different velocities. This important concept is considered in more detail at a later point. It is the purpose at this point to identify examples of interviews that make use of manipulative aids.

Interviews Relating to Number (classes, series, and so on)

Experiments during the 1930s and early 1940s resulted in a book, Piaget and Szeminska (1941a), which summarized the results of a great many interviews with children. The first two interviews described in the book have since become famous and are widely used by workers in the field to test for conservation of discontinuous and continuous substance.

Discontinuous substance or liquids were used to determine whether children of various ages could establish and maintain equivalence of a given volume of liquid. Two identical glasses were filled to the same point with a liquid (see Figure 1–2).

After the child agreed that both containers held the same amount of liquid, the contents of one glass were poured into a much larger container so as to cause the water level to be relatively lower. The child was then asked if the amounts of liquids were still equal. This procedure was repeated with other containers, always being careful not to mix the two initial liquids. Liquid *A*, for example might be poured into five smaller containers and then compared with liquid *B*.

For discontinuous quantities, beads were used with containers similar to those used for continuous substance. In addition, however, the beads could be strung together and length was used to compare quanti-

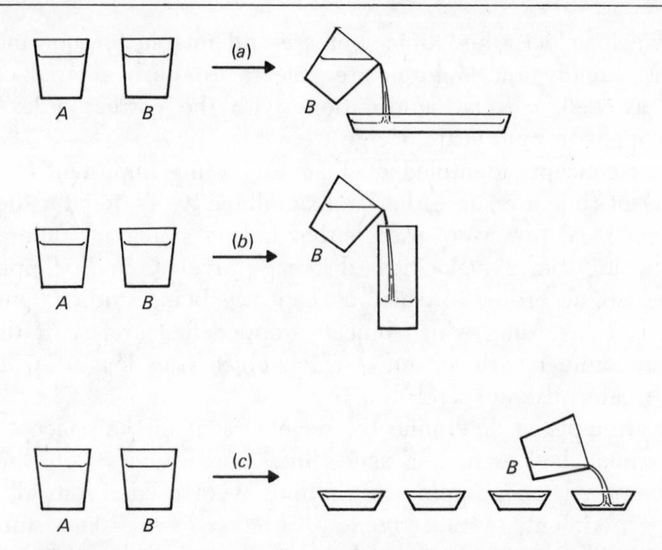

Figure 1–2. Apparatus for conservation of continuous substance.

ties. The child could be asked if the length of the necklaces are the same, referring to the two sets of beads in the specified containers. A further variation in the interview procedure was that of establishing one-to-one correspondence between the two identical quantities of beads. Each time the interviewer placed a bead in a container, the child also placed a bead in his container. When all beads were in both containers, the interviewer would then make variations in one or the other and ask the child to compare the amount of beads.

To investigate seriation and ordinal correspondence abilities, tests were devised that included sets of objects that could be ordered according to length, width, diameter, and the like. Two sets of objects might be compared and elements of the series interchanged to determine the seriation or ordering abilities of the children. Additive and multiplicative composition of classes were studied in some detail in the 1940 book and in much greater detail with a large ($N > 2000$) sample as reported in a later work, Inhelder and Piaget (1959). In this book outlining the early growth of logic, the development of classification ability is reported in detail. Objects of various sizes, shapes, colors, and the like were used in interviews with children aged two to nine years.

A recent study by Good et al. (1973) should serve to illustrate interviewing technique, materials, and criteria for response determination by children. One hundred children in first and second grade were interviewed individually using materials known as "people pieces" and developed by the Elementary Science Study (ESS). The set of people

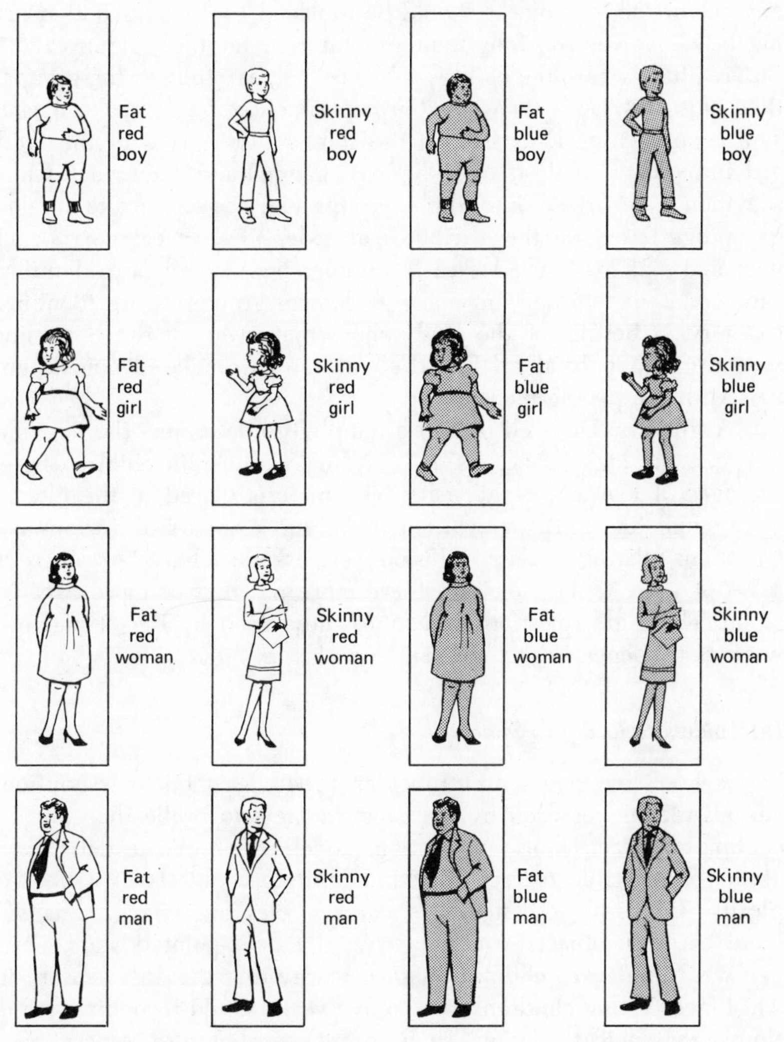

Figure 1–3. Sets of people pieces.

pieces consists of 16 plastic squares, each of which has a figure depicting a person with certain characteristics. Four girls, four boys, four women, and four men constitute the set with the other attributes being red, blue, skinny, and fat (see Figure 1–3).

The experimenter *E* places the 16 pieces on the table in front of the child and places the fat red man, fat red woman, fat red boy, and skinny red girl into a group. The experimenter *E* then asks the child if any one of the four pieces should be replaced, and why. Another grouping task involves the four families (fat red, fat blue, skinny red, and skinny blue) where the child is asked to look carefully at the groups and then close his/her eyes while the experimenter *might* make a change. The experimenter *E* exchanges the skinny blue girl with the fat blue girl and asks the child if anything was changed and if so, what. Children are then categorized into three groups. Those who make no logical groupings based on the attributes are identified as category 3. Those who make at least one logical grouping, but not all, are identified as category 2, and those who make all logical groupings are identified as category 1. Results of the study show that even by the beginning of second grade, only about half the children can make all logical groupings with the people pieces.

A study by Dodwell (1962) attempted to determine the relationship between the logic of classes and cardinal number in children between the ages of five and eight years. The materials used in the interviews were small toy dolls, toy garden tools, and toy cars of various colors. Questions relating to class inclusion were asked such as: "Are there more rakes or more tools?", or "Are there more red cars or more cars?", and so on. These are questions similar to those used by Piaget and his co-workers at Geneva in experiments on classes.

Interviews Relating to Space

A book by Piaget and Inhelder (1948a) contains descriptions of various techniques used by the experimenters to probe the child's conception of space. Haptic perception (touch) was used to determine the ability of the child to recognize and reproduce objects of various complexity. Simple objects, such as squares, circles, and triangles, as well as more complex objects, including swastikas, six-pointed stars, and the like were used with children in such a way that the only sensory input was touch. Young children (two to five years) could recognize and draw simple topological relations such as openness, closure, separation, and so on, but were unable to coordinate angles and lines involved with Euclidean relations.

Linear and circular order were studied using strings and beads of different colors, "surrounding" was studied using knot analysis and tying, projective geometry concepts were probed using objects and their shadows, and many other spatial concepts were investigated using various concrete aids. More details of the nature and results of these and other experiments with the child's conception of space are given later.

Interviews Relating to Deductive Logic

Although a great many other areas of reasoning have been studied and could be used in this section as examples of interviews using manipulative aids, they are deferred to a later section when the results of cognitive development research are analyzed in some detail. The final examples of interviews using manipulative aids can be found in a book by Inhelder and Piaget (1955), which presents experiments designed to probe the development of logico-deductive thinking in seven to 15-year-old children. Only a few of the 15 or so experiments described in the book are presented here to further illustrate interviews involving manipulative aids.

1. The Billiard Apparatus

   The child's ability to discover and generalize the principle of equality between angles of incidence and reflection was probed using apparatus similar to a billiard table. Balls were launched from a plunger that could be directed in various directions. Figure 1–4 shows targets (circled numbers) that were placed at different locations on the table.

   The interviewer would ask the child to aim the rebound for the various targets and various questions were asked to encourage the child to explain why the ball behaved as it did. This experiment was designed to assess the child's ability to deal with reciprocal implications; that is, the angle of incidence necessarily implies the angle of reflection and vice-versa. Not until the age of eleven to twelve years do children discover this relationship between angles of incidence and reflection and not until about the age of fourteen to sixteen years do the subjects reflect true hypothetico-deductive thinking.

2. Floating and Sinking Apparatus

   Various pieces of wood, metal in various forms (foil, needles, keys, balls, and the like), pebbles, candles, corks, and the like

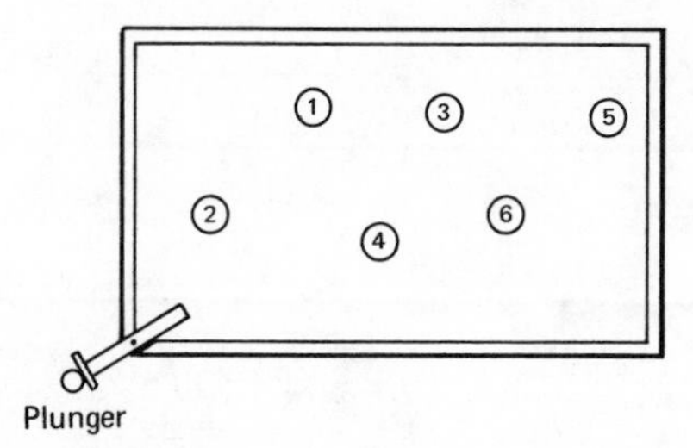

Figure 1–4. Billiard table apparatus.

were used with buckets of water to determine classification schemes and generalizations of objects that float or sink. Since the law or principle to be discovered involves the density of water as compared to that of other objects, conservation of weight and volume are necessary conditions for a complete understanding of the experiments. The interviewer asks the child to classify the objects according to whether they will float or sink. The basis for each classification is then asked and finally the child is given buckets of water to test the predictions. A discussion follows this experimentation in which the interviewer encourages the child to generalize the ideas and findings into a single conclusion or "law." The child's ability to construct appropriate classifications, revise them according to the experiments, and eliminate contradictions that arise is exposed by this interview procedure. It was found by Inhelder and Piaget that only toward the age of thirteen to fourteen years did the children give indications that they could discover systematically the density relationships among the objects with respect to the density of water.

3. Pendulum Apparatus

Identifying and controlling variables in an experiment are critical factors in investigations involving natural phenomena. Figure 1–5 shows the apparatus used by Inhelder and Piaget, in collaboration with Morf et al., to investigate the child's ability to control variables.

The fairly obvious variables involve the weight of the object, the length of the string, the height of the dropping point, and the magnitude of push given to the object. To find the degree of influence of each of the variables on the time for each complete swing, one must change only one of the variables at a time while

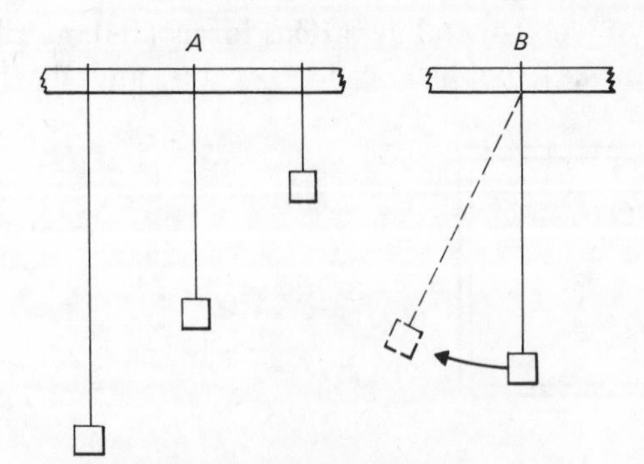

Figure 1–5. Pendulum apparatus.

Figure 1–6. Liquids for forming combinations.

holding all others constant. Again, it is not until the age of fourteen to fifteen years that children take a systematic approach to separating out the variables and then controlling them one by one.

4. Apparatus for Forming Combinations

The final example of interviews using manipulative materials is one used to test for the development of a complete combinatorial system. Figure 1–6 shows the apparatus involved.

Five different colorless liquids are used: (1) dilute sulfuric acid, (2) water, (3) oxygenated water, (4) thiosulfate, and (5) potassium iodide. When liquids 1, 3, and 5 are combined the mixture turns yellow. Adding liquid 2 to the mixture will do nothing but adding liquid 4 will cause the color to disappear.

The child is presented a flask containing liquids 1 and 3 and another flask containing liquid 2. The interviewer then pours a small amount of liquid 5 in each of the two flasks. The child is then asked to reproduce the yellow color using the five flasks 1–5 in any way he/she wishes. Consistent with the other experiments already presented in this section, only children of the age of fourteen to fifteen years old could approach this problem in a systematic manner.

Scores of other experiments have been devised for the purpose of interviewing children and many of them are described, with results and conclusions, in the remaining parts of the book. It is hoped that at this point the reader has a "feeling" for the nature of some of the interview procedures used to collect data on cognitive development. Interviews with manipulative materials are obviously superior to purely verbal interviews and the great majority of studies presented in Part Two deals with such research.

references

Dodwell, P. C. "Relations between the Understanding of the Logic of Classes and of Cardinal Number in Children", *Canadian Journal of Psychology,* Vol. 16, 1963, pp. 152–60.

Good, Ronald G., Charles Matthews et al. "Cognitive Development Characteristics of Children in Grades 1–5." A part of a comprehensive study presented at the 46th Annual Meeting of the *National Association for Research in Science Teaching,* Detroit, 1973.

Piaget, Jean. *Le langage et la pensée chez l' enfant.* Neuchâtel: Delachaux and Niestle, 1923; translated by M. Gabain, *The Language and Thought of the Child.* London: Routledge and Kegan Paul, 1926.

———. *Le jugement et le raisonnement chez l' enfant.* Neuchâtel: Delachaux and Niestle, 1924; translated by M. Warden, *Judgment and Reasoning in the Child.* New York: Harcourt Brace Jovanovich, 1926.

———. *La représentation du monde chez l' enfant.* Paris: Librairie F. Alcan, 1926; translated by J. and A. Tomlinson, *The Child's Conception of the World.* New York: Harcourt Brace Jovanovich, 1929.

———. *La causalité physique chez l' enfant.* Paris: Librairie F. Alcan, 1927; translated by M. Gabain, *The Child's Conception of Physical Causality.* Totowa, N.J.: Littlefield, Adams, 1960.

———. *Le jugement moral chez l' enfant.* Paris: Librairie F. Alcan, 1923; translated by M. Gabain, *The Moral Judgment of the Child.* New York: Harcourt Brace Jovanovich, 1932.

———. *La naissance de l' intelligence chez l' enfant.* Neuchâtel: Delachaux and Niestle, 1936(a); translated by M. Cook, *The Origins of Intelligence in Children.* New York: International Universities Press, 1952.

———. *La construction du riel chex l' enfant.* Neuchâtel: Delachaux and Niestle, 1936(b); translated by M. Cook; *The Construction of Reality in the Child.* New York: Basic Books, 1954.

———, and Alina Szeminska. *La genese du nombre chez l' enfant.* Neuchâtel: Delachaux et Niestle, 1941a. English translation, *The Child's Conception of Number,* New York: Norton, 1965.

———, and Barbel Inhelder. *Le développement des quantities chez l' enfant.* Neuchâtel: Delachaux et Niestle, 1941b.

———. *Le dévelopement de la notion de temps chez l' enfant.* Paris: Presses Universitaires de France, 1946(a); translated by A. Pomerans, *The Child's Conception of Time.* London: Routledge and Kegan Paul, 1969.

———. *Les notions de movement et de vitesse chez l' enfant.* Paris: Presses Universitaires de France, 1946(b); translated by G. Holloway and M. Mackenzie, *The Child's Conception of Movement and Speed.* London: Routledge and Kegan Paul, 1970.

———, and Barbel Inhelder. *La représentation de l' espace chez l' enfant.* Paris: Presses Universitaires de France, 1948a; translated by F. Langdon and J. Lunzer, *The Child's Conception of Space*. London: Routledge and Kegan Paul, 1956.

———, Barbel Inhelder, and Alina Szeminska. *La géométrie spontanee de l' enfant*. Paris: Presses Universitaires de France, 1948b; translated by E. Lunzer, *The Child's Conception of Geometry*. London: Routledge and Kegan Paul, 1960.

———, and Barbel Inhelder. *De la logique de l' enfant à la logique de l' adolescent*. Paris: Presses Universitaires de France, 1955; translated by A. Parsons and S. Seagrin, *The Growth of Logical Thinking from Childhood to Adolescence*. New York: Basic Books, 1958.

———, and Barbel Inhelder. *La Genèse des structures logiques elementaires*. Neuchâtel: Delachaux and Niestle, 1959; translated by E. Lunzer and D. Papert, *The Early Growth of Logic in the Child*. London: Routledge and Kegan Paul, 1964.

———. *Biologie et connaissance*. Bussière, France: Editions Gallimard, 1967.

———, and Barbel Inhelder. *Mémoire et intelligence*. Paris: Presses Universitaires de France, 1968.

part two

SCIENCE-RELATED CONCEPTS AND THEIR DEVELOPMENT

objectives

It is intended that this section will help the reader to

1. Identify specific intellectual characteristics of persons who are at a "preformal" level of development.
2. Determine ways of finding out what an individual's developmental level is with respect to such abilities as classification, seriation, number operations, and the like.
3. Carry out individual interviews with students in an attempt to determine their "concrete" conceptual levels.
4. Become familiar with research evidence available on the thinking abilities of preformal persons.
5. Identify specific intellectual characteristics of persons who are at a "formal" level of development.
6. Determine ways of finding out what an individual's developmental level is with respect to such abilities as identifying and controlling variables, using proportions, using deductive logic, and the like.
7. Carry out individual interviews with students in an attempt to determine their "formal" conceptual levels.
8. Become familiar with research evidence available on the "formal" thinking abilities of individuals.

Objectives 5 to 8 deal with the abstract or formal thinking abilities of individuals whose ages are usually beyond eleven to twelve years. For the reader who is primarily interested in children of elementary school age, the material that explains probability, combinations, proportions, and other formal operational abilities can be dealt with very briefly, if at all.

An understanding of the nature of the process of cognitive development in children and adolescents is needed if one is to make rational decisions about helping students learn science. If one's ability to learn is directly correlated to developmental growth, then providing appropriate learning opportunities depends upon an understanding of that person's level of development. The nature of the curriculum and how the student is introduced to the curriculum (teacher role) should be based on an understanding of the students in any given classroom.

introduction

Most concepts and principles in the domain commonly called "science" are actually combinations of concepts or thought processes that can be identified in cognitive development research. The various science textbook series and curriculum development projects that have appeared during the past 15 years generally do not reflect what is currently known about the thinking abilities of children. The purpose of this part of the book is to identify thought processes that are closely related to the development of science concepts. Time, for example, is obviously a very important science concept that underlies all motion studies. Research that throws light on how children gradually develop an operational understanding of time can be of considerable help to those in education who try to take a thinking approach to curriculum development and instruction.

The ability to identify and control variables is a prerequisite to any type of systematic experimentation with science-type systems. Even the simplest problems requiring an investigative approach to possible solutions, usually involve at least three or four variables. Something apparently considered relatively easy according to placement in many science curricula is growing plants from seeds. It does not take much searching to realize that a great many variables might have some effect on the growth of the plant. Seed type, soil type, position of the seed, temperature, and amount of moisture represent dozens of possibilities that might affect the sprouting of the seed. The extent to which the child can conceptualize all of the possibilities and systematize the approach to each possible solution is dependent on that individual's cognitive development level.

It should be emphasized here that although isolated thought processes will be considered and related to corresponding science concepts and principles, the human mind develops as a complex interrelated set of structures. Particularly at the onset of what Piaget calls concrete operational thought, around the age of seven or eight years, the child's intellect develops as a complex whole, with one mental operation being linked to others.

Part Three briefly considers the "theory" developed by Piaget as a result of considering the voluminous amount of research and, at that time, the specific concepts and processes are presented as more of an integrated whole reflecting more accurately the nature of developing thought processes.

concept development at a concrete level

According to the research conducted and theory developed by Piaget and his co-workers at Geneva, the human mind does not show evidence of true logical thinking until the age of eleven or twelve years and even then it is only the beginning of the stage of "formal" operations. "True" logical thinking must, of course, be defined operationally before one can meaningfully discuss the onset of formal operations (logical-thinking). This is done in some detail later in Part Two, but a brief description here should help to distinguish prelogical from logical thinking.

Inhelder and Piaget (1955) have identified the most distinctive feature of formal or logical thought as being able to formulate totally abstract possibilities relative to that which is empirically measurable. In doing this, one must use the operations that are common to deductive logic: conjunction, disjunction, implication, and equivalence along with their reciprocals and negations in a completely interrelated way. The resulting complete combinatorial system allows the individual to go beyond the concrete, physical environment to a consideration of abstract possibilities, not necessarily tied to anything measurable in the "real" world. This system of "16 binary operations" and the transformational group known as the INRC group are considered in some detail later and related to science principles or systems requiring logical thinking.

Prelogical and concrete thinking as considered in Part Two are limited largely to the ages of four or five to eleven or twelve years. This range corresponds roughly to the elementary school levels, K–5 or 6. Among the concepts and principles involved with prelogical and concrete concept development and related to "science-type" thinking are classes, series, ordinal and cardinal number, length-area-volume relationships, spatial coordination, weight and mass, density, time, motion, equilibrium, geometry, chance or probability, combinations, and general ideas associated with physical causality. The various science concepts, principles, conceptual schemes, and the like that are often found in textbook series and some curriculum projects such as matter-energy conservation, atomic-molecular nature of matter, cellular-genetic nature of life, and the like are considered at times in conjunction with research on cognitive development functions, particularly in the later section on abstract thinking.

Classification

The development of the ability to classify objects and events in increasingly more complex ways is seen as highly important and central to the attainment of formal or logical thinking. Studies on classification ability during the late 1930s and early 1940s by the Geneva group are summarized in *The Child's Conception of Number* by Piaget. An even more comprehensive and detailed work involving the examination of over 2,000 children appeared in 1959, *The Early Growth of Logic in the Child* by Inhelder and Piaget. Much of the information that appears in this part on classification is drawn from these two sources.

Perceptual and spatial factors are central to the early attempts by the child to analyze objects in terms of identifiable attributes. In order to classify objects by shape, one must be able to identify similarities and differences. Until about the age of three to four years, children's perceptual space is limited to simple topological relations such as surrounding, openness, and closure. The number of sides of closed curves (square, pentagon, and the like) are not copied correctly until the age of four to five years and then only simple Euclidean shapes can be represented accurately. Not until the age of seven to eight years can most children begin to distinguish among fairly complex shapes and represent them through drawings. Although the ability to form classes does not derive from perception alone, it is a prerequisite for forming classes.

Preclass Collections

Genuine class making involves the coordination of two kinds of properties or relations. *Intensive* properties are those that are common to the elements of a certain class. It is these properties that are used to determine how a class is to be formed. For example, Figure 2–1 shows various objects that have been grouped together in a class according to the two *intensive* properties: (1) each figure must be three sided, and (2) each figure must be closed.

The second property or relation necessary to constitute a class is *extension.* Extensive properties are those that relate one class with others. For example, dogs belong to the larger class of animals. Coordination of intensive and extensive properties allows for hierarchical classes as shown in Figure 2–2.

Until intensive and extensive properties are established and coordinated, genuine classification is not realized. Before the age of seven to eight years children make collections of objects that are tied to the

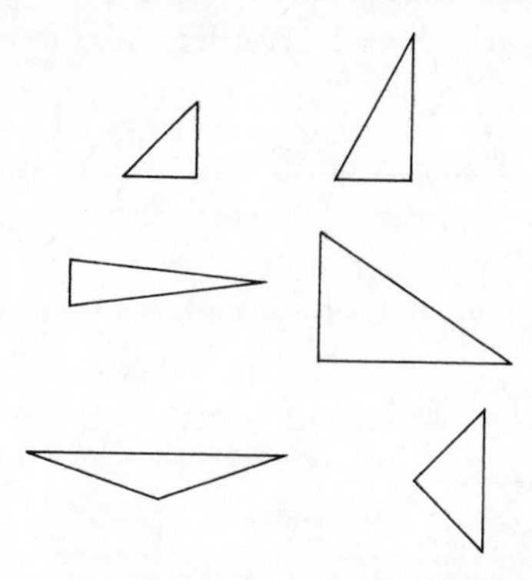

Figure 2–1. Example of a class formed by two *intensive* properties.

spatial configuration of the objects, relative to one another. The objects in the child's environment are a determining force in placing objects together. Until about the age of five years, children place things together not because they are alike in terms of intensive properties but because they form familiar objects or perhaps just because they form a spatial configuration that is pleasing. The triangle must be placed atop the square because a house is formed. Objects are placed in a particular arrangement that results in a familiar "picture." Sometimes objects are placed in alignment according to successive similarities as in Figure 2–3.

Similarities are determined by spatial or temporal factors and quite often three- or four-year-olds will change the criteria for alignment many times. It is often seen of four- and five-year-olds that they will complete an alignment where the criteria have fluctuated and then go back and rearrange a few elements. This is an indication that the child is beginning to notice intensive properties and to group accordingly. However, it will be another two years or so before intensive and extensive properties are coordinated to form hierarchical classes.

One-dimensional alignments evolve into two-dimensional alignments

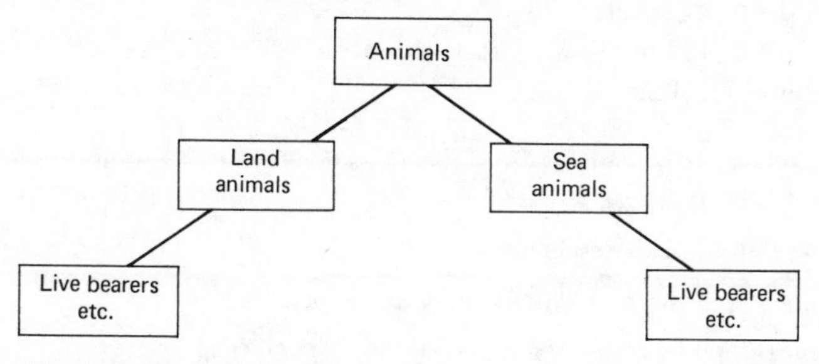

Figure 2–2. Coordinating intensive and extensive properties allows for hierarchical classes.

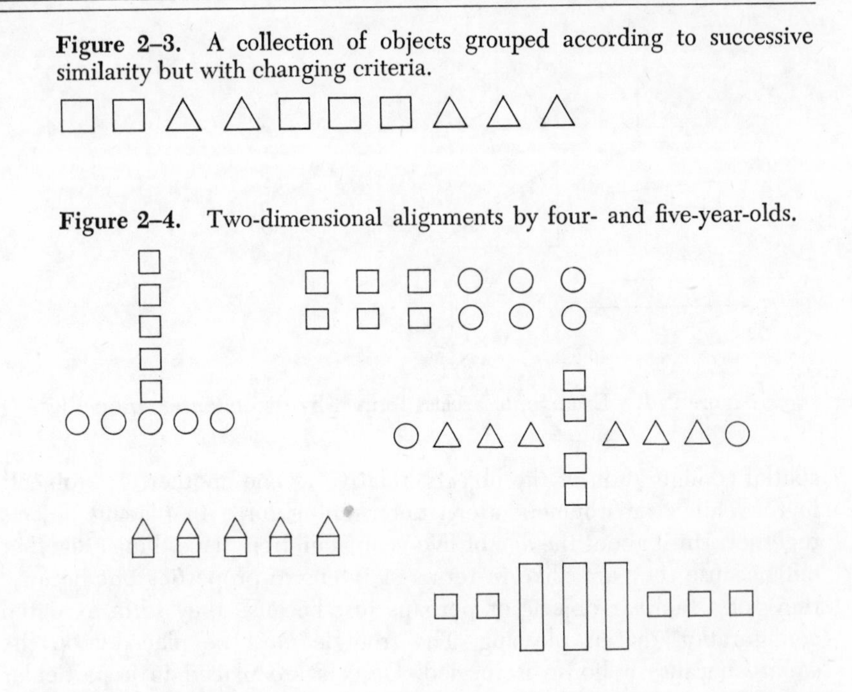
Figure 2–3. A collection of objects grouped according to successive similarity but with changing criteria.

Figure 2–4. Two-dimensional alignments by four- and five-year-olds.

as the child continues in this preclassification stage. Figure 2–4 shows examples of two-dimensional collections.

It should be pointed out that merely because a child of say eight years forms two-dimensional alignments, it is not a sign of inability to form real classes. He/she may just feel like making such shapes. This is a very different situation from the four- or five-year-old who is *unable* to make genuine classes.

Between the ages of four or five to seven or eight years, children become more consistent in using *intensive* properties but are as yet unable to relate *extensive* properties to form hierarchical classes. Spatial configuration is becoming less of an influential factor in determining why things should go together.

Inhelder and Piaget (1959) have identified ten criteria required for logical or genuine classification:

1. All elements or objects must be classified even if a unique element must occupy a class by itself.
2. There are no isolated classes.
3. A class includes *all* elements having a specified property.
4. A class includes *only* elements having a specified property.
5. All classes of the same rank are disjoint.

6. A complementary class has its own characteristics, not possessed by its complement.
7. A class is included in every higher ranking class that contains all its elements.
8. The inclusion of elements in a class are the minimum compatible with intensive properties.
9. Similar criteria distinguish classes of the same rank.
10. In symmetrical subdivision of a class, if a class B_1 is subdivided into A_1 and A_1' and the same criterion is applicable to B_2, then B_2, must be subdivided into A_2 and A_2'.

These 10 criteria or characteristics of true classification are used by Piaget to differentiate classes from collections as seen in the actions of children. The main problem confronting children in the construction of classes is the coordination of extension and intension. Development of the ability to coordinate extensive and intensive and relations properties of elements of classes is now viewed in terms of "class inclusion."

As defined previously, extension involves thinking of a class as a union of elements, whereas intension refers to each element representative of certain attributes. A given attribute will form part of the intensive properties of a class if *all* members of the class have this attribute. To know whether an object should become part of a subclass within a larger class requires the conceptualization of all and some. Figure 2–5 shows sets of objects that could be used to determine a child's understanding of the relationship between all and some.

Children are asked to divide a set of the objects into two, four, or more groups and are asked questions according to what groupings occur. For the circles and squares, such questions as: Are all the circles blue?; Are all the red ones square?; Are all the blue ones circles?; Are all the squares red?, and so on would be used to determine the child's concept of all and some. For the set of objects that are all rectangles in Figure 2–5, more subclasses can be formed but similar questions can be asked to determine class inclusion ability. Usually not until the age of seven to eight years can children demonstrate adequately that they understand the concepts of all and some, and even then many have not yet developed

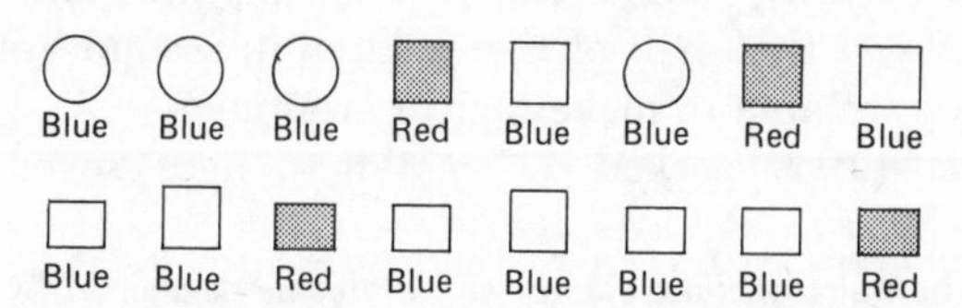

Figure 2–5. Objects used to test for "all"–"some" conditions.

the understanding $B = A + A'$ or $A = B - A'$ or $A' = B - A$. They have not quantified the difference between all and some such that all *must* refer to B whereas some can only refer to either A or A'.

An example at this point of simple classification as compared to hierarchical classes might help to show the difficulty children have in making logical classes. Good, et al. (1973) interviewed 48 first grade and 49 second grade children on a classification task using people figures printed on plastic squares.[1] Figure 1–3 depicts the attributes of the "people" that differ in height (or age), width, sex, and color.

The experimenter *E* placed 16 people pieces on a table and grouped the fat red man, fat red woman, fat red boy, and skinny red girl together. *E* then asked the subject S if any one of the group should be replaced by one of the remaining 12 pieces and why? If the response was "fat red girl" and S indicated a logical reason, *E* went on to another task. If the response was something different, *E* would try to find the reasoning behind the choice.

Other tasks using these materials presented to the S, including a game where four families (fat red, skinny red, fat blue, skinny blue) are grouped and S was asked to study the four groups. *E* then asked S to look away and S was told by *E* that a change might take place. In fact, *E* exchanged the skinny blue girl with the fat blue girl and S was then asked if anything was changed and if so, to "change them back."

In scoring the interviews, children who gave all correct answers and logical reasons were labeled as category 1. Category 2 included children who gave at least one correct choice with a logical reason. Category 3 children were those who were unable to give logical reasons for any tasks. Reasons were seen as logical if the child could give some indication that the attributes of the people pieces were being used in a consistent way to make the groupings.

Table 2–1 shows the results of the 97 individual interviews. The interviews were conducted during the second month of the 1972–1973 school year and reveal considerable differences in the classification abilities of first and second graders.

Less than one fifth of the children in first grade could make logical groupings on all three parts of the task, whereas almost one-half of the children in second grade were able to make logical groupings. This study assessed what appear to be fairly simple classification abilities and yet the results indicate that fewer than half of the children in second grade, early in the school year, were able to make logical groupings.

Inhelder and Piaget (1959) asked 117 children, ages seven to

[1] "People pieces" from the unit *Attribute Games and Problems* developed by the Elementary Science Study.

TABLE 2–1. Classification Task: Number and Percentage of Children Who Were in Category 1, 2 or 3

Grade	Category 1 (All Logical Groupings)	Category 2 (At Least One Logical Grouping)	Category 3 (No Logical Grouping)
1	9 (19%)	25 (52%)	14 (29%)
2	23 (46%)	16 (32%)	10 (22%)

thirteen years, questions about the classification of animals. Three or four sketches of ducks, and sketches of an equal number of other birds comprised part of one set. There were also sketches of a fish, a horse, a dog, a snake, and a mouse in the set. In another set of cards, sketches of three ducks, four other birds, a bee, a butterfly, a dragonfly, a bat, and three inanimate objects were used to determine the class inclusion and hierarchical classification abilities of the children. Each child was asked to make two piles, four piles, and so on using the two sets of pictures and the interviewer would ask questions such as: Are the ducks animals?; Are the birds animals?; If all the animals were killed, would there be any feathered animals left?; Are there more birds or more animals here?; Where should the fish go?; Are there more ducks or more domestic animals in the world?; and so on. All of the questions are aimed at probing the child's ability with class inclusion and hierarchical classification. Table 2–2 shows the percentage of correct responses to the various questions.

Interestingly enough, a third of the twelve to thirteen-year-old group have not yet developed the ability to make logically consistent class inclusions. This means that a significant number of children in upper elementary and beginning middle or junior high schools are quite limited in their ability to form logical classes and subclasses or sets and subsets.

TABLE 2–2. Percentage of Correct Answers to Questions on Class Inclusion of Animals

Ages	8	9	10	11	12–13
ducks < animals	38	66	62	82	75
birds < animals	38	66	62	82	75
both	25	27	42	46	67

A study by Kofsky (1966) attempted to look at the order in which classification ability develops. A basic assumption underlying Piaget's theory of cognitive development is that various concepts are acquired by children in a fixed order. Kofsky interviewed 122 children, ages four to nine years, to determine whether Inhelder and Piaget's conclusions (1959) about the ordered development of classification ability would be confirmed. Geometric blocks 1-inch thick with a plane surface area of about 4 square inches, of different colors were used to test the following classification abilities of the children:

1. Resemblance sorting (*RS*)—grouping two objects together that look alike.
2. Consistent sorting (*CS*)—grouping more than two objects together based on similar attributes.
3. Exhaustive sorting (*EC*)—grouping together all things that possess a particular attribute.
4. Conservation (*CON*)—spatial arrangement does not influence classes.
5. Multiple class membership (*MM*)—elements of a set might also belong to other classes if a "reclassification" occurred.
6. Horizontal reclassification (*HR*)—criteria for grouping elements of a set are changed to allow for a different grouping.
7. Hierarchical classification (*VC*)—attributes are combined into combinations in constructing successive classes.
8. Some and all (*SA*)—objects can belong to more than one class and subsets can be formed from a more inclusive set.
9. Part-whole relationship ($A + A'$)—similar to SA, but now the words *some* and *all* are used in joining subsets to form a larger set.
10. Hierarchy conservation ($B - A'$)—the reverse of classification nine where a large class is divided into constituent parts.
11. Inclusion (*BA*)—a realization that a class identified by certain criteria is always larger than its subclass, that is, animals horses.

These 11 steps were analyzed for each of the 122 children, and Table 2–3 from the Kofsky study shows the results.

There is a relatively high correlation (.87) between the predicted sequence of class development and the sequence shown in Table 2–3. The variation from a perfect correlation might be explained in many

TABLE 2–3. **Percentage of Subjects in Each Age Group Passing Each Task**

Task	Age Group (Years) 4	5	6	7	8	9	Total
CS	90	90	95	100	100	100	96
RS	81	95	85	100	95	100	94
EC	43	75	80	90	90	90	77
SA	52	50	90	90	85	95	77
MM	10	35	60	90	95	100	65
$A + A'$	29	40	65	81	75	90	62
$B - A'$	14	20	45	66	80	95	55
CON	29	35	60	43	45	75	45
HR	10	25	20	66	70	90	45
$B > A$	29	20	10	19	45	60	31
VC	10	25	0	14	20	40	17

ways but the important outcome is the obvious differences in difficulty of various classification tasks using the same objects for all tasks.

Many other studies have been conducted in an effort to learn more about the abilities of children of various ages to form classes. Annett (1959) found that 60 per cent of the children younger than eight years changed their criteria for forming classes if more than two objects were involved. The nature of the objects involved would surely be an influential factor in such a study, but an important implication is clear. Even simple classes are not possible if class criteria are changed when three or more objects are introduced.

In another study supporting the evidence and theory considered thus far, Blair Hood (1962) found that most children younger than seven years believed that the superordinate or largest possible class was actually smaller than the larger of two subordinate classes.

The importance of the development of a complete classification system has long been recognized and is central to Piaget's theory of the development of logical thinking. Inhelder and Piaget (1955) described in great detail the importance of classification in logical thought development. The logic of classes and relations was described as the central means for problem solving during the six- or seven to eleven- or twelve-year old range. During this time the child is developing the mental structures necessary for the development of formal logic or abstract thought. Much more is said of the development of formal or abstract thinking later in Part Two, but it should be recognized by the reader at this time that classification is central to the problem of the analysis of cognitive development in children of elementary school age.

Seriation

Seriation or the process of making a series with a group of objects is, like classification, a basic cognitive process in the development of logical thinking. Seriation differs from classification in that the relations between the elements can be perceived as longer, shorter, thinner, heavier, and so on, whereas a class, as such, is not as easily subject to perception. A series of sticks that are ordered from shortest to longest is perceptually the "best form." However, operational seriation is not based on perception alone since children are unable to deal with simple and multiple series until after about seven or eight years of age. This is long after they are capable of discriminating between lengths, widths, and so on.

In one study by Piaget and Morf (1958) children aged four to ten years were interviewed with sets of sticks that had a regular variation in length from one stick to the next. Some had equal variations from one stick to the next so that the series went up in a straight line whereas others varied such that the tops of the ordered sticks formed one half of a parabola. Among other things, it was found that before the ages of nine or ten years, children are unable to refer to the shape of the line across the top of the series of sticks in comparing differences in pairs of adjacent sticks. Instead, they must measure one pair against the other to determine the relative differences. This suggests that the younger children do not use the shape of the entire series in solving problems within the series.

An earlier study by Piaget and Szeminska (1941) used 10 small rods ranging from 9 to 16 centimeters together with a set of rods that could be inserted between each of the rods (see Figure 2–6). Thus, when combined properly the set consisted of 18 ordered rods. Only by about eight years of age could the children begin to systematically order the 10 rods from shortest to longest or vice-versa, and then insert the other 8 rods in the proper places. Younger children were sometimes successful in ordering the 10-rod set and inserting the other 8 rods but only through a trial and error process. This indicates a lack of conceptualization of the process of seriation.

The age at which children generally become operational with respect to seriation depends upon the nature of the materials involved. Seriation of weight, for example, develops about two years later than seriation of length. Conservation studies reported later in Part Two point up the variation in the development of concepts that are commonly seen as "science-type" concepts.

The experiments on seriation reported by Inhelder and Piaget

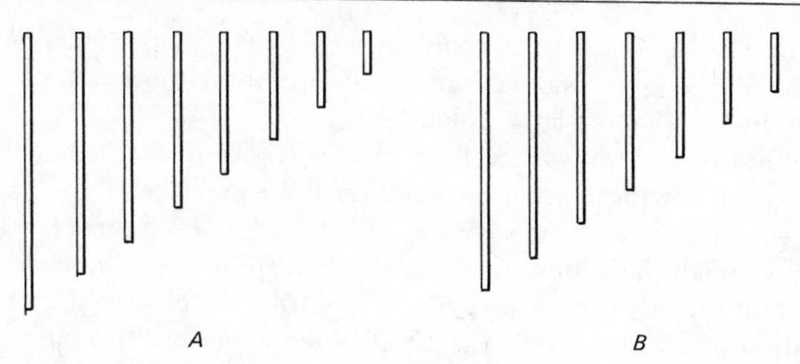

Figure 2–6. An ordered set of rods and an accompanying set for insertion.

(1959) include two parts to the interview. First, the child is given 10 rods differing in length by 0.8 centimeters or 8 centimeters from smallest to largest and each having its own color. They are in random order, and the child is asked to draw the set as it will look when it is ordered. After the drawing, the child is then allowed to order the rods. The results for 88 children, aged four to nine years, are shown in Table 2–4.

Rather surprisingly, the data show that children of four to six years of age are better able to anticipate the correct order of the rods by drawing them than they are in actually ordering them. At first thought this would appear to contradict a logical developmental sequence whereby children are able to do progressively more difficult tasks. Piaget, however, explains the problem in terms of a lack of reversibility on the part of five- and six-year-olds that is needed in operational seriation but not in drawing the correct series. Systematic seriation requires that a person be able to compare a given element of the series with both shorter and longer elements. Drawing the series requires only that one line be drawn after another in *one* direction. Thus, drawing requires only a one-directional, nonreversible process whereas actual seriation

TABLE 2–4. **Anticipation (Drawing) and Performance in Seriation (Per Cent)**

Age **Number of Subjects**	**4** **19**	**5** **33**	**6** **19**	**7** **10**	**8–9** **7**
1. Failure in anticipation	89	42	5	0	0
Global anticipation	11	55	73	20	0
Analytic anticipation	0	3	22	80	100
2. Failure in seriation	84	54	42	0	0
Success by trial-error	16	40	36	20	14
Operational seriation	0	6	22	80	86

requires a two-directional, reversible process. The importance of reversibility of thoughts and actions is discussed in more detail with conservation experiments later in Part Two.

Seriation or ordering as well as classification are studied in more detail in the following section on number. Both processes develop in a parallel way and as has been strongly postulated by the Genevan school, primarily through the actions of the child. Perception and language play an important but not the most critical role in the development of either classification or seriation. It is the child's own actions with objects and events in his immediate environment that tend to promote the development of both of these critical cognitive processes.

NUMBER

Experiments in science nearly always involve attempts to quantify results in terms of number concepts. It is, therefore, important to consider the development of the concept of number in children before we pursue a study of the development of other science-type concepts. The previous look at classification and seriation should prove helpful in understanding how the child gradually conceptualizes the rather abstract idea of number.

Both classification and seriation are completely basic and central to the development of one's conception of number. Number involves relations between sets and it has already been established that classification is a process of grouping elements into sets whereas seriation is a process of relating one element of a set to another, as in ordering from shortest to longest. Two sets *X* and *Y* are equal if for each element of *X* there is an equivalent element in *Y* and vice-versa. The set of whole numbers from 1 to 10 involves both the concept of seriation and that of classification.

Piaget and Szeminska (1941) followed the development of the concept of number in children from the ages of four to eight years, and later Inhelder and Piaget (1959) extended the work with children up to twelve to thirteen years of age. Number concepts as used here should not be confused with merely "number names," counting, or even the arithmetic skills of adding, subtracting, and the like. All of these can be accomplished merely through rote learning or memorization with little understanding for the underlying concepts involved. Children can learn to count to 100 by the time they are five years old but it is not a truly operational kind of counting. The cognitive activity is similar to learning to repeat the letters of the alphabet from *A* to *Z*. Counting in an operational sense presupposes an understanding of seriation or order and the ability to progress in either direction within a series. We have already seen that before the age of seven to eight years few children are capable of genuine seriation, so counting for the five-year-old is necessarily a rote process, probably done to please an adult.

Two ideas that are basic to an operational understanding of number are one-to-one correspondence and conservation. One of the now well-known "conservation" experiments developed by Piaget and his co-workers at Geneva involves a test for number conservation. As a prerequisite to the task, the child is asked to establish one-to-one correspondence between two sets of objects. An adaptation of the original number task was used by Good et al. (1973) and is described here with results and discussion.

The experimenter (*E*) makes a row of eight black checkers and asks the subject (*S*) to make a row of red checkers such that there is one red checker for each black one (see Figure 2–7).

If S is unable initially to establish one-to-one correspondence between the red and the black, *E* might ask S to make a pile with the red checkers and then try the "game" again. If it becomes evident that S is unable to establish one-to-one correspondence, then the task is terminated, since the conservation part of the task remaining requires that the S understand the initial equivalence of each set of checkers. When S establishes one-to-one correspondence, *E* then moves all of the black checkers into a pile and asks if there are more red, more black, or the same number, and why? *E* then makes a row with the black checkers about twice as long as the row of red checkers and repeats the questions. If the child answers correctly to all questions and gives logical reasons he/she is considered to be a conserver of number on this task and is labeled as category 1. If S gives at least one correct answer with a logical reason, he/she is considered transitional and is labeled as category 2. If there are no logical reasons with correct answers, the child is considered a nonconserver and is labeled category 3. In grades 1–5, 249 children were interviewed individually on this task and the results are shown in Table 2–5.

Notice that before the age of seven to eight years (first grade) a majority of the children (53 per cent) are unable to conserve or hold in their minds the equivalency of the two sets of objects even though they

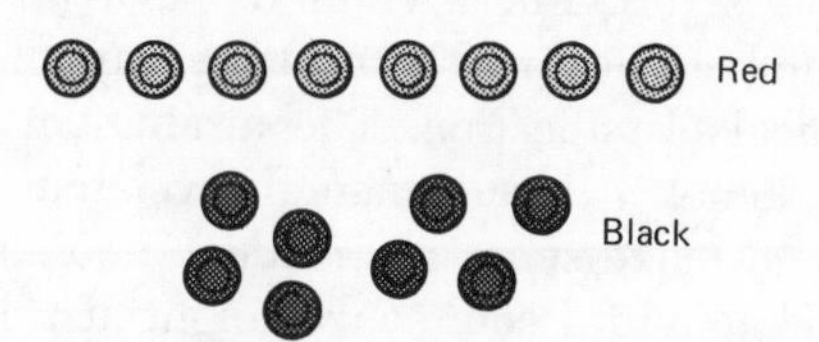

Figure 2–7. Materials and arrangement for number conservation interview.

TABLE 2–5. **Results from the Conservation of Number Task**

Grade	Category 1 (conserver)	Category 2 (transitional)	Category 3 (nonconserver)
1	23 (47%)	12 (24%)	14 (29%)
2	35 (73%)	9 (19%)	4 (9%)
3	39 (78%)	9 (18%)	2 (4%)
4	48 (96%)	1 (2%)	2 (2%)
5	50 (98%)	1 (2%)	0

established initially the one-to-one correspondence. Merely changing the spatial arrangement of one of the sets of checkers caused the children in categories 2 and 3 to change their minds and say that one of the sets now contained more checkers than the other. During a period of about seven to nine years of age, approximately three-fourths of the Ss have attained conservation on the task and beyond that almost all of the children are in category 1.

These results compare favorably with those of Piaget and others who have studied various aspects of number concept development but by no means do they explain the relative complexity of how a child attains an operational understanding of number.

A number such as 7 or 18 or 135, or the like refers simultaneously to quantity or cardinality and to a place in a series or ordinality. The cardinal value of 7 refers to a quantity of elements in a set, whereas the ordinal value of 7 refers to the position of an element in a series. The process of cardination involves class operations or grouping quantities of elements together based on selected criteria. For instance, one might want to group all prime numbers less than 100 together: 1, 3, 5, 7, 11, 13, 17, and so on for a particular purpose. The resulting set of whole numbers constitutes a class generated by a certain criterion. The process of generating the class produces a series of numbers 1, 3, 5, 7, 11———97. Let us look at a single number for a moment to assess its simultaneous nature of cardination and ordination. The number eleven denoted by the symbol 11 suggests both relational and class characteristics. It is at once a quantity and a place in a series. For a child to have a truly operational understanding of number, he or she must be capable of reversible seriation and hierarchical classification. The attainment and coordination of these operations signal a developmental level that will only then allow a child to begin to make sense of arithmetic computations and other mathematical-type activities commonly taught at a level of difficulty beyond the conceptual level of many children.

One experiment used by Piaget (1941) to test for the interdepend-

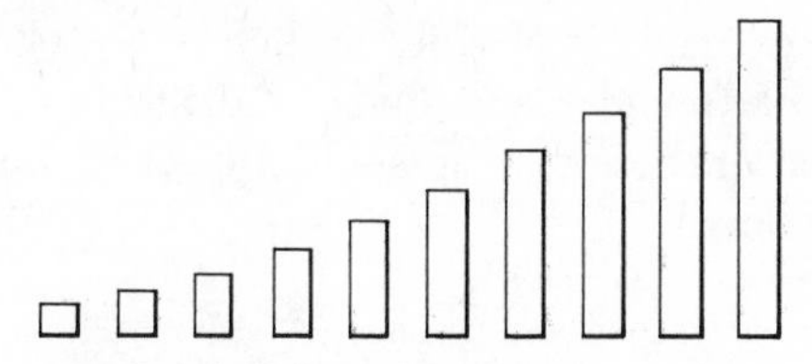

Figure 2–8. Materials for assessing interdependence of ordination and cardination.

ence of seriation and cardination used an ordered series of 10 cardboard pieces. Figure 2–8 shows the relative dimensions of the pieces.

The child is asked to make a staircase with the pieces, count them beginning with *A*, and then to say how many pieces like *A* could be made from *B* or *C* or *H*, and so on. The interviewer (*I*) then picks a piece at random from the staircase and the child is asked how many units (like *A*) the piece represents. If the child at this point can make the cardinal value (for *H* it would be 8) correspond with the ordinal value (*H* is eighth in the series or staircase) it is an indication that he or she understands the relationship between cardination and ordination. If the child must measure for each determination of "how many units," however, this is an indication that the relationship has not been grasped. Once the position of an element in such a series is known, the magnitude or cardinal value is also known. The sixth element is always 6 units, the seventh is always 7 units, and so on. Such a series of objects obviously corresponds to the series of whole numbers 1———*n*.

The results of this and other similar experiments show that ordination and cardination develop simultaneously and it is not until the age of seven to eight years that children develop the understanding of this simultaneous relationship. Before the reader jumps to the conclusion that all children beginning at the age of seven to eight years can be expected to understand addition, subtraction, multiplication, and division with whole numbers as well as with fractions, decimals, and the like, other things must first be considered. Such concepts as singular class, null class, all-some relations, double classification, multiple seriation, conjunction or intersection, disjunction or union, left-right relations, and "place value" are also involved and necessary for a functional understanding of mathematics at a given level of difficulty.[2] We attempt a brief look at some of these concepts to emphasize the relative complexity of number concept development.

Most of the number-related concepts just identified do not develop

[2] Refer to Appendix *B* for paper entitled "The Traditional Sequencing of Mathematics for Young Children Should Be Changed."

in a fully operational way for the majority of children before the age of ten to eleven years. The work of Inhelder and Piaget (1959) reported data on the development of most of these concepts and a brief summary of the procedures and results follow.

Singular Class

Up to the ages of eight or nine years, children have difficulty in recognizing or constructing a set containing a single element. The materials used in the interviews consisted of three to six triangles and one of them had a cross on one side. This triangle was always a different color from the others so even though the cross was on the hidden face, the observer could always tell which triangle had the cross. The child was always asked for a reason for his/her choice and then asked to make up a similar puzzle. For children from the age of five to seven years there was only a 50 per cent level of success on choices and either partial or no apparent understanding. At the age of seven to nine years the success level rose to about 75 per cent with understanding of the "puzzle" system. When the children discovered the "rule" or pattern that the cross is always on the triangle that appears different, they give reasons which indicate that class inclusion thought is being used.

Null Class

The null class or empty set is similar to the singular class in terms of difficulty for children. Neither a set with only one element nor a set without any elements coincides with a child's idea of a class as a collection. Children resist the idea of a null set until about the age of ten or eleven years or very close to the onset of "formal" operations. This is because the period of "concrete" operations, roughly seven to eleven or twelve years, is based on real objects in the child's environment, but the null class or empty set is void of elements. Thus, to make sense of the null class, one must be able to abstract the notion of a class without elements.

The experiment reported in *The Early Growth of Logic* to assess the child's conception of null class involved square, round, and triangular cards, some with pictures of trees, houses, and the like, whereas others were blank. The child was asked to group the cards in any way and then place them into only two groups. The logical grouping would be blank cards versus nonblank cards. Not until the age of ten or eleven years do the children finally group the cards into blank and nonblank.

All-Some Relations

The importance of class inclusion in hierarchical classification has been considered in the section on classification and so will be treated only briefly here. A set (all) and a subset (some) are used repeatedly in "modern" mathematics as a means of introducing more "understanding" in elementary mathematics and less rote learning. Unfortunately, many children are introduced to set theory before they have developed the mental structures to allow them to make sense of the various operations at that level of difficulty.

The basic interview technique in all-some experiments is to ask the child if all of *X* are *Y*, that is, are all the squares, red? Experiments using geometric shapes, weights and balance, flowers, and the like all confirm the fact that it is not until after the age of eight or nine years that all-some relations are understood by most children.

Multiple Classification

For double classification (two criteria determining the class) children can complete correctly many tasks by the time they are seven or eight years. For three criteria, however, it is not until after the age of nine or ten years that children classify consistently. This is about the same age at which the intersection of classes or "cross" multiplication is achieved. Cross multiplication of classes was assessed by Inhelder and Piaget (1959) using materials similar to those shown in Figure 2–9. The child is asked what object should be placed in the square (S) where the groups of objects intersect. To be sure that each child has carefully

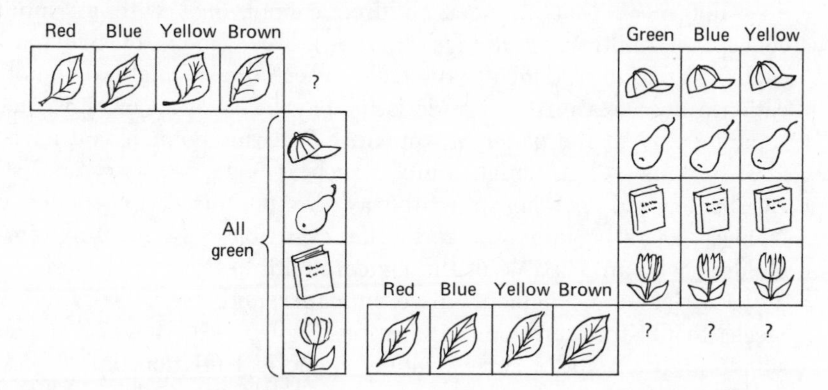

Figure 2–9. Materials used to assess intersection or simple multiplication.

considered each row of objects, the child is asked such questions as: Why have all of these been placed together? How are they similar or different? Up to the age of seven to eight years most of the children interviewed made a choice for the open square based on only one of the sets of objects rather than on both collections. The object chosen is often not represented in either of the sets of objects even though it may bear a strong similarity. For example, a tree may be used instead of a leaf. When the children begin to choose an object as a function of both sets, there is still a gradual progression toward operational multiplication. They may zero in on only one element from each set rather than taking all of them into account. Or, at a later point in their development, they may take all objects into account (say leaves of different colors in one set and all green objects in the other) but still choose the wrong object, that is, a green object but not a green leaf. By about the age of nine or ten years roughly one half of the children were able to place the most logical object at the intersection of the two sets.

The similarity between the logical multiplication of classes and the process of cross multiplication of sets (represented by numerals) should be fairly obvious. One potential major difference is that a procedure for computing answers using numbers can be learned by rote without necessarily understanding the meaning or implications of the procedure. This, in fact, is quite often the case since children in third grade (eight to nine years) are taught a procedure called multiplication that, according to the research findings we have just considered on multiple classification, roughly one half of the children may as yet be unable to comprehend the meaning upon which the procedure is based. Even the children who are capable of logical multiplication or intersection of classes of objects are still dependent upon concrete objects for their thinking. This does not mean that younger children cannot deal with an intuitive concept of multiple groupings (that is, three groups of two marbles each). The level of difficulty typically expected of formal cross-multiplication procedures in third grade is far beyond such an intuitive understanding. (Read the paper in Appendix *B* at this point if you have not already done so.) Although symbols such as 3, 14, 53, *X*, =, and so on can be learned in a meaningful way through direct experience with "things," there is no assurance of this even for eight- to nine-year-old children who are capable of the logical multiplication of classes.

Matrices are commonly used in paper and pencil activities as a means to illustrate cross multiplication. Figure 2–10 shows examples of simple matrices and a corresponding numerical relationship.

The procedure of manipulating numerals is learned by children as multiplication even though little real meaning may be attached to the process by the children. To relate meaning to the process of this symbol

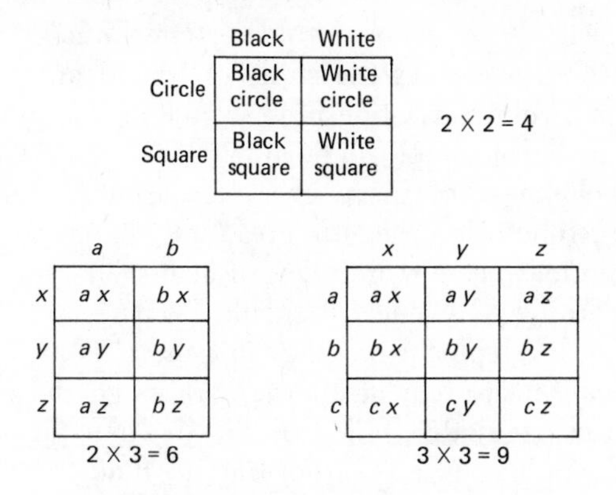

Figure 2–10. Matrices as cross multiplication.

manipulation, the child must be able to construct and complete corresponding matrices.

Arithmetic operations derive from the conversion of logical elements or objects, such as leaves, marbles, and stones, into iterable units that allow for more precision. Three is obtained by iterating the unit 1, three times, and so on. One can look at the learning of arithmetic operations as a process of changing actions with objects into symbols with greater possibility for quantification. Measurement operations also deal with iterable units (12 inches/foot, 100 centimeters/meter) and are studied in some detail in later sections on time, geometry, and distance and motion.

Much more could be said about the development of the concept of number in children and certainly a great many studies attest to the importance of this concept. Hopefully, at this point the reader has a more detailed conceptualization of the concept of number as it develops in children. In general, this development is based on the coordinated development of seriation and classification and is not a simple matter of learning number names, counting, adding, subtracting, and other procedures that can be based wholly on rote memorization.

Space Concepts

Piaget and Inhelder (1948a) have referred to the study of the concept of space as an indispensable part of child psychology. Their joint work published in 1948, *The Child's Conception of Space,* considers the development of the concept of space from a general sense. Children from the ages of three to twelve years were studied in terms of topological, projective, and Euclidean space. Another volume, *The Child's Conception of Geometry* (1948b), continues the study of Euclidean space and in particular length-area-volume relationships as well as coordinate systems. We defer the study of geometry until after a brief but representative look at the development of space in a more general sense. Although the measurement functions in length, area, and volume and a workable coordinate system may seem more relevant at first in studying concepts related to the learning of science, it should become apparent that a brief analysis of the development of space in general is, in fact, basic to the study of specific measurement abilities. Although the core of what is presented in the pages that follow is drawn from the two works already cited, other studies are used from time to time for comparison and breadth.

Topological Space

Piaget and Inhelder (1948a) found that young children develop concepts related to topological space before they develop the more quantitative concepts of Euclidean geometry. The concepts of proximity or closeness, separation, order, and enclosure or surrounding are developed before concepts of angles, distance, and shape rigidity. Topology, or the study of what some have called "rubber sheet" geometry, developed historically after Euclidean geometry even though from a cognitive development standpoint it precedes Euclidean concepts. The idea of rubber-sheet geometry means that shapes are equivalent topologically as long as the rubber sheet is not torn or separated. Thus, a square is equivalent to a circle, a cube to a sphere, a drinking glass to a swimming pool, and so on. Figure 2–11 shows examples of closed figures that are equivalent in topology.

The rubber sheet can be stretched in all of these examples without any tearing or separating. Angles, distances, sides, and the like are not factors used in determining if two figures are topologically equivalent. Figure 2–12 shows examples of figures that are not equivalent topologically.

The development of the basic topological concepts of proximity, separation, order, surrounding, and continuity were studied by Piaget by asking children to make certain drawings (proximity and separation), string beads (order), tie knots (surrounding), and to subdivide lines and figures (continuity and infinity). Through these and other related studies, data were collected that suggest how children are able to represent topological space at various points in the continuum of development. A summary of the methods and findings is presented here.

Free or spontaneous drawings by children from the ages of three to

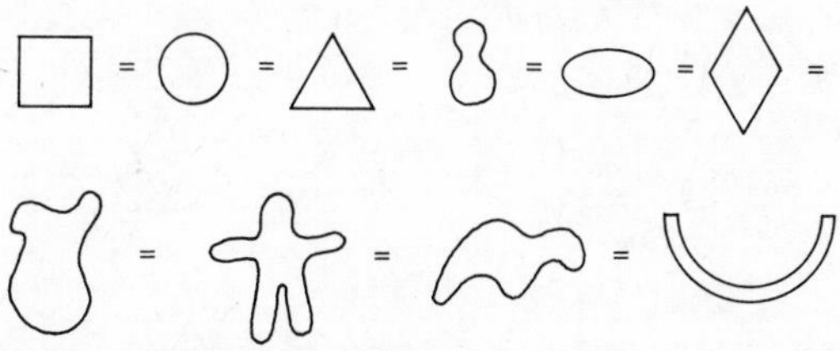

Figure 2–11. Closed figures topologically equivalent.

nine years were analyzed. Those who have watched the drawings by children take shape will recognize many of the findings. Children up to the age of four to five years do not attend closely to the Euclidean space relations of proportions and distances but the topological relations of proximity and separation are clearly attended to even by the 4-year-old. In drawing a face, the eyes, nose, and mouth are fairly close to each other, yet still separate. It is not uncommon, however, to see the nose below the mouth. The "order" of these parts has not yet been developed in terms of representation even though proximity and separation are clearly evidenced. Surrounding or enclosure is also largely lacking during Stage 1. Evidence for this conclusion is seen in the same drawings of children who place the eyes outside the face or a chimney under the roof.

From the age of four or five to the age of seven or eight years there is a gradual progression in sophistication of the child's ability to represent the topological relations of proximity, separation, order, enclosure, and continuity.

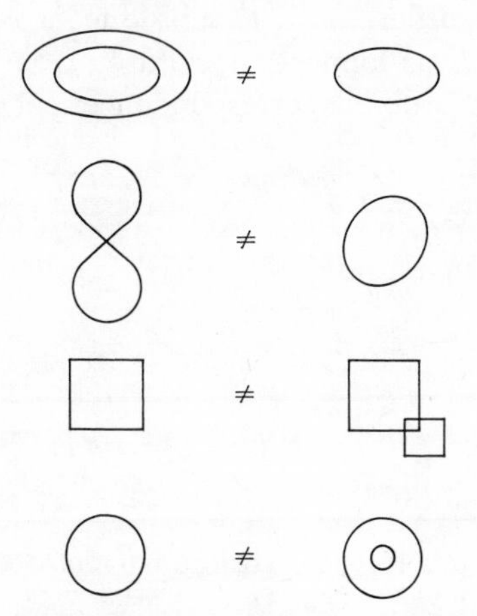

Figure 2–12. Figures not topologically equivalent.

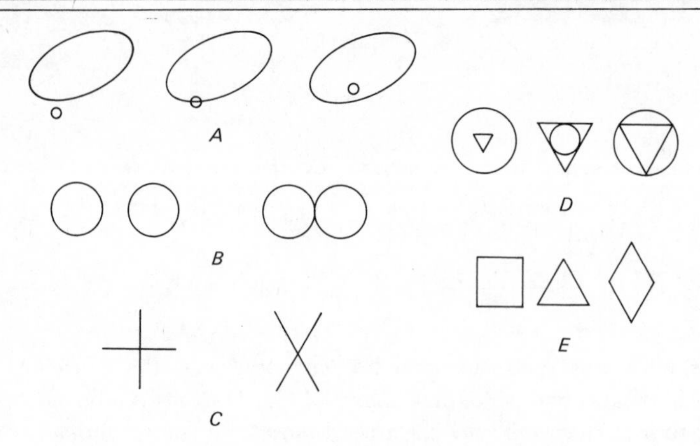

Figure 2–13. Shapes used to determine drawing and construction abilities.

Figure 2–13 shows examples of shapes used to determine the child's ability to draw and to construct with matchsticks, some of the shapes.

It is found that only by the age of six or seven years are most children able to reconstruct the rhombus (diamond-shape) and the circumscribed figures to any degree of accuracy in terms of Euclidean properties. The topological relations are represented earlier than Euclidean relations as seen in the drawings of the various shapes. Figure 2–14 shows examples of drawings that are topologically correct but incorrect when viewed from Euclidean relations.

Linear and circular order were investigated by asking children to do various "stringing" tasks with colored beads. Among the tasks were: (1) make a string of beads identical to a model; (2) transpose circular into linear order; (3) establish reverse order; and (4) reproduce a string of beads in a figure-8 arrangement. It was found that children from the ages of four to six years could make ordered correspondences, at first

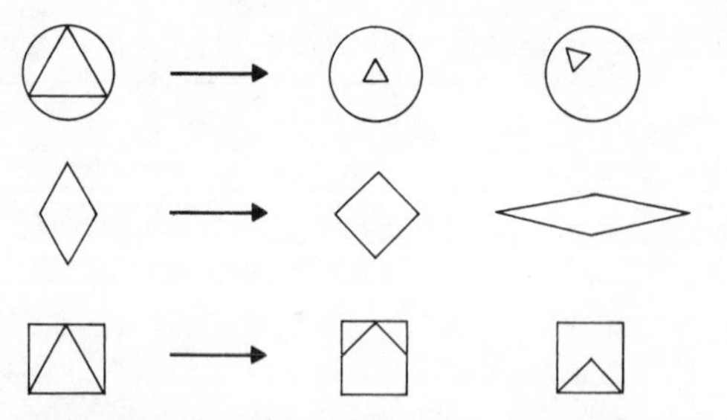

Figure 2–14. Drawings that are topologically correct but not according to Euclidean relations.

only when the model is in the field of vision and later, to some degree, when the model is not in the same field of vision. Circular order can be transposed into linear order near the age of six, but a reversed order is still not possible, except with simple figures. By the age of seven years most children are able to deal both with linear and circular order in an operational way although reversal of order with more complex figures such as a figure 8 still pose difficulty for many children.

The development of the relationship of surrounding was studied by asking children to tie various knots shown loosely tied or, if needed, shown being formed. All knots were fairly simple such as the left- and right-hand clovers shown in Figure 2–15.

Up to the age of four years few children were found to be capable of tying knots and could not identify a "false" knot. From about the ages of four to seven or eight years the children developed greater ability to tie knots and correctly identify false knots, although it was not until the age of six or seven years that they could relate taut and loose knots or identify false knots with any regularity. After the age of seven or eight years most children seem to be able to copy simple knots by tracing the loops and see that a false loop, which is not actually intertwined, will not form a knot when the ends of the string are pulled. During the period of from seven to eight to eleven to twelve years, children develop the ability to substitute imaginary lines for strings in a continuing development of organizing the relations of proximity, separation, order, and surrounding into an organized, workable whole. It is to this end that the concepts of points, continuity, and infinity are studied.

Continuity is an extremely important and widely pervasive concept in science as well as in mathematics. Conceptualizing physical and biological processes as continuous in nature is necessary for any genuine understanding. Such concepts and processes as temperature and its measurement, motion and its measurement, growth and its measurement, and chemical change and its measurement are continuous and divisible beyond our ability to perceive the change in individual units, for example, in quantum units for energy changes. The importance of this concept of continuity cannot be overemphasized in the learning of science concepts and principles that require a type of abstraction in-

Figure 2–15. Left- and right-hand clovers.

herent in an operational grasp of continuity and limits. The process of subdividing a line indefinitely is not conceptualized as possible until the age of eleven or twelve and even then many children are rather unsure.

Among the techniques used by Piaget and Inhelder (1948a) in assessing the developmental nature of continuity were: (1) asking children to draw the smallest and largest squares possible; (2) subdividing a closed figure (circle, square) as many times as possible; and (3) asking a child if a line or surface could be a series of points and then asking a child to illustrate, using paper and pencil. Until the age of seven or eight years the results of the interviews showed that children are unable to conceptualize infinite subdivision or continuity. From that point to the age of eleven to twelve years there is a gradual development in the ability to conceptualize infinite subdivision or infinite change of any kind. It is emphasized that continuity concepts are still tied to the child's concrete environment and that it is not until after the age of twelve or thirteen years that continuity can be imagined apart from what is real and experiential.

One rather interesting implication for mathematics education is the commonly used definition in elementary school of a line or surface as an infinite set of points. It would appear that such a definition is beyond the majority of children until the age of eleven or twelve years and, of course, there is no guarantee that understanding will suddenly appear "on schedule" even after this age.

The study of topological relations (topology) has greatly increased during the twentieth century, and certain aspects of this type of "space study" such as order and continuity have obvious relatedness to science concepts. However, precisely because the relations are independent of any contractions or expansions of figures, they are of little value in measurement activities such as those involved in space-time studies. In this respect, topological space relations are more primitive than those of projective and Euclidean space where points are located according to perspective or projective systems and coordinate systems.

A well-known study by Dodwell (1963) reported results that suggest strongly that there may not be as clear-cut a progression of steps as suggested by Piaget and Inhelder in the development of spatial concepts in children. Dodwell's study does not conflict with the essence of the results in this section on topological space or those that follow in other sections on projective and Euclidean geometry, but it serves to point up the relative complexity of the development of cognitive structures related to spatial concepts. The ages reported in any study at which children develop certain abilities should be interpreted only as approximate ages. Many children in a given group might not correspond to the suggested age ranges for the development of spatial concepts, number concepts, and the like, but this does not disprove the theory of cognitive development by Piaget or the evidence upon which this theory is based.

Projective Space

The study of projective space is sometimes referred to as "shadow" geometry. Drawing or imagining straight lines and being able to view objects from various perspectives are required in projective space, whereas they were not required in topological space. A square is no longer equal to a circle as is the case in topology. Relations involved in projective geometry can be illustrated with a light source, objects, and a screen as in Figure 2–16.

The shadow of the object is seen on the screen and will vary according to the object and its orientation. The size of the object is unimportant in projective geometry, just as in topology. A solid cylinder placed end-to-end between the light source and the screen will appear as a solid circle or a disc on the screen. If it is turned so that its axis is vertical (parallel with the screen), the shadow that appears on the screen is that of a rectangle. The fundamental ability in determining the shadow shapes is that of drawing or imagining straight lines from the source to the various points on the object and finally to the screen. If you are directly behind the object, that is in place of the light, it is fairly easy to predict the shape of the projected shadow, but from other positions it becomes less easy. Conceptual development of the straight line and the further development of coordinated perspective were investigated by Piaget and Inhelder (1948) using various simple techniques.

For the straight line, a circular table and a square or rectangular table are required along with a quantity (15 to 20) of matchsticks, each stuck in a small base of plasticene. The first task consists of having the child make a straight line with the sticks between the two end posts or sticks that have already been placed by the interviewer. At first, the two end posts are placed parallel to and near one edge of the table and later they are placed at an angle of 30 or 40 degrees to the edge. With the round table, the end posts are placed so that a straight line would form either the diameter or a chord. As a final task, the child is shown a series of posts in zigzag fashion and asked to put them into a straight line. The results show that up to the age of about seven years children either cannot construct a straight line except along one edge of the rectangular table or they must use lengthy trial-and-error procedures to complete the task successfully. From about the age of seven years the child learns to "sight" along the posts to form the straight line and does not need successive trial and error any longer.

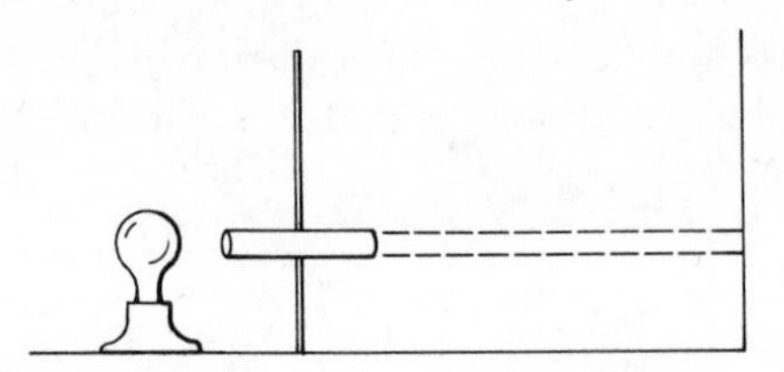

Figure 2–16. Materials for projected shadows.

The basic requirement for forming different perspectives is to imagine, not merely perceive, straight lines projecting in any direction. Development of a coordinated perspective was assessed by Piaget and Inhelder in a number of ways. One was to show a child various objects such as a thin rod, a disc, and other objects that are geometrically simple but which will appear different from various perspectives. The child is seated at a table, and a doll is used to ask questions about how things will look from various points around the table. Each child is asked to make a drawing of how the object will look at each viewpoint and later is shown some drawings and is asked to choose the correct one. At about the age of six or seven years the children begin to show signs of developing a coordinated perspective, although it is still far from complete and operational. By about the age of eight or nine years most children exhibit the ability to view simple objects from various viewpoints, but for more complicated (irregular) objects another two to three years is required for the children to develop further coordination of perspective.

In experiments with the projection of shadows using objects of constant cross section (pencil, rectangle, disc) and objects of varying cross section (single and double cones), Inhelder and Piaget found that coordination of perspective for cones was more difficult. Children by the age of eight or nine years could predict correctly the shadow of objects at various orientations, but with constant cross section, whereas two or three years more were required for them to make successful prediction with cones.

Figure 2–17 shows the well-known "mountains task" used by Piaget and Inhelder to assess the coordination of perspective in children from the ages of four to twelve years.

It is similar to a previous experiment in that a table, chairs, a doll, and pictures of the three mountains are used to determine the child's ability to change perspective mentally according to the position of the doll. The reverse of this was also done, with the child's being shown a picture from among 10 and asked where the doll would have to be to have such a view. By the age of 7 or 8 years, the children began to show signs

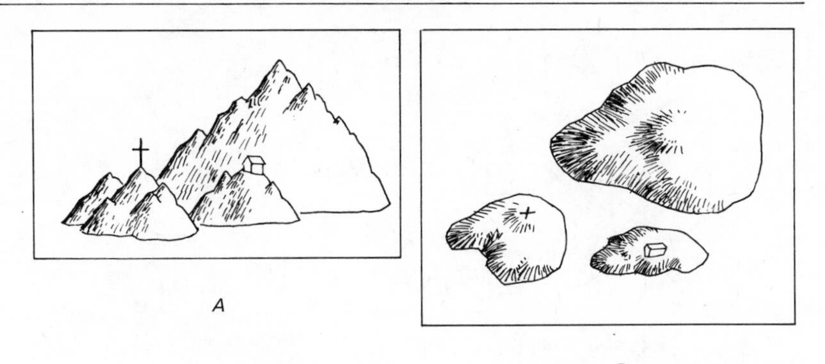

Figure 2–17. Mountains task used for coordination of perspective.

of separating the different points of view but with little success. From that point until the age of eleven or twelve years, the children showed a progressive coordination of perspective, with fairly good success at the age of nine or ten years. The study referred to earlier by Dodwell (1963) replicated the mountains task and achieved essentially the same results. A more recent study by McNair and Slack (1972) again points up the danger in interpreting results in a rigid all-encompassing manner. With subjects in tenth grade, they found that four out of nine subjects were unable to correctly coordinate all four perspectives in the mountains scene (front, back, left, right). Although the number of students is rather small, it serves to point up once again the pitfalls in interpreting research results in a nonflexible manner or in overgeneralizing the results. It is useful to talk of stages of development, but it is also important to remember that there is no "average" child. There are only statistical averages and they should be interpreted with all their inherent limitations, which are many. It has also been shown that the nature of the interview situation and criteria used for interpreting the data are critical factors in the variation of results with many experiments (Good et al., 1973; Minichillo and Goodnow, 1969; Bittner, 1968). These variations are analyzed in more detail later in Part Two.

In a study related to the coordination of perspective, Good (1972) interviewed 60 children in grades K–4 to determine their ability to construct three-dimensional structures using cubes and a sketch of the block structure. Since the sketch offered only one viewpoint and in most cases "hid" some blocks from view (see Figure 2–18), the child had to imagine the structure from various viewpoints. The child was shown a sketch drawn on a sheet of paper with the exact dimensions of the cubes (two centimeters on a side), which were later used by the child in attempting to build the structure. The interviewer first asked the

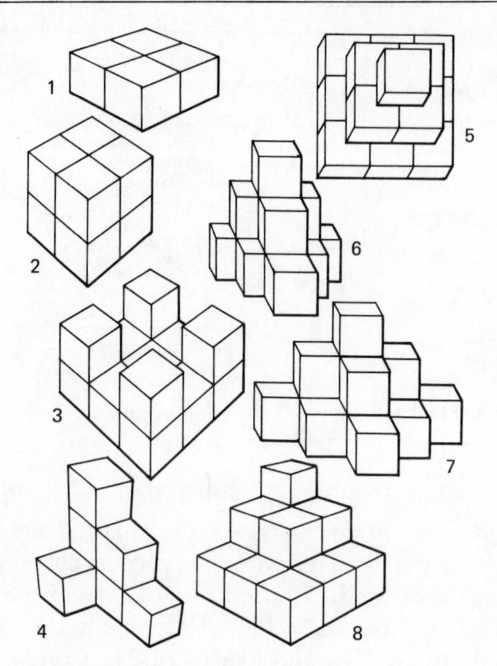

Figure 2–18. Block structures.

child how many blocks would be necessary to build the structure. After the response, each child was given the number of cubes requested and was asked to build the structure represented by the sketch. The sketch remained in front of the child during the construction process, and when the child was satisfied with his or her efforts, the interviewer would repeat the procedure using another sketch. The time required for each construction was recorded without the child's awareness and the child could receive as many blocks as were deemed necessary. Table 2–6

TABLE 2–6. **Percentages of Correct Number of Block Requests**

Grade	Figure 1	2	3	4	5	6	7	8
K	83	33	17	0	0	8	0	8
1	92	67	0	8	0	0	0	8
2	83	67	17	33	17	17	8	25
3	100	92	83	58	42	17	0	8
4	92	75	58	67	50	0	25	50

shows the percentages of correct requests, and Table 2–7 shows the percentages of successful completion of block structures. In comparing the results shown in Tables 2–6 and 2–7, it can be seen that children were generally less successful in predicting the correct number of blocks required than in actually building the structures. This is not surprising since in the construction of some of the figures, children would often realize that they needed one or two more blocks to complete the structure. Their original estimate through visual perspective was made more accurate through actual manipulation. The data in Table 2–7 actually are related more closely to the coordination of perspective as we have been dealing with it since the construction of the block structures sometimes forced children into realizing that their original estimate was in error. Structure number 6 posed the greatest degree of difficulty for the children followed by structures 7, 8, and 5. Structure 6 most often was built similarly to structure 7, with the children apparently having great difficulty with only portions of each block visible in the sketch. Consistent with this, Table 2–6 shows that no children in grade 4 were able to request the correct number of blocks required for the construction of structure number 6, and only one third were able to complete a correct construction. This particular structure also required more time on the average (63 seconds) than the others; 45 seconds for structure 7; 48 seconds for structure 8; 40 seconds for structure 5.

A coordinated perspective would allow the child to imagine each sketch from various viewpoints and thus allow the individual to "see" what might be hidden in the sketch. There is a wide range of success on the various structures for all age levels, from the age of five or six years to the age of nine or ten years with major difficulty on sketch 6. These results reinforce what was mentioned previously about the variation in data being caused by the nature of the interview materials, among other things. A coordinated perspective is not "complete" at a prespecified time for any given individual but depends upon experi-

TABLE 2–7. **Percentages of Successful Completion of Block Structures**

	Figure							
Grade	**1**	**2**	**3**	**4**	**5**	**6**	**7**	**8**
K	75	50	17	33	17	8	0	17
1	83	66	33	58	17	0	25	17
2	83	83	83	92	42	0	50	67
3	100	100	67	100	75	50	42	58
4	100	100	83	100	83	33	42	67

ences even after the eleven- or twelve-year point when Piaget suggests that all subjects should have developed such a coordinated perspective. Completeness suggests no further development, and such a static view of cognitive development is inconsistent in terms of available data as well as a logical analysis of general cognitive development theory. A more reasonable view of spatial development, as well as of logical development in general, is that the rate of development is much slower after a given stage, but nonetheless continues. As progressively more difficult situations are encountered, mental structures continue to develop and reach a stable but dynamic equilibrium.

In two brief but highly interesting sections in *The Child's Conception of Space,* Inhelder and Piaget present findings on geometrical sections and the rotation and development of surfaces. In the geometric sections experiments, objects such as cylinders, cones, and prisms were made of plasticene and presented to a child who was then asked to predict the shape of the surface produced if the solid were to be cut along various planes. Before the cut was made, the child was asked to draw the surface and then to pick out a drawing from among many. Figure 2–19 shows objects and their plane surfaces after cutting.

For fairly simple objects such as the cylinder and cone, the children begin showing some facility with the cuts that are either exactly parallel or perpendicular to the axes of the objects by the age of six or seven years. As the cut is "angled" from the axes, the children have more difficulty. With more complex objects such as a corkscrew shape, it is not until the age of eleven or twelve years that children begin to solve the problems satisfactorily.

In the section on rotation and development of surfaces, an attempt was made by Inhelder and Piaget to assess the development of the child to mentally rotate and develop a cylinder, cube, pyramid, and cone. Each child was presented with the solids and asked what these objects would look like if they were "unfolded." Through the child's drawings of each unfolded shape, the experimenters were able to see a gradual development of the ability of children to mentally reconstruct solid

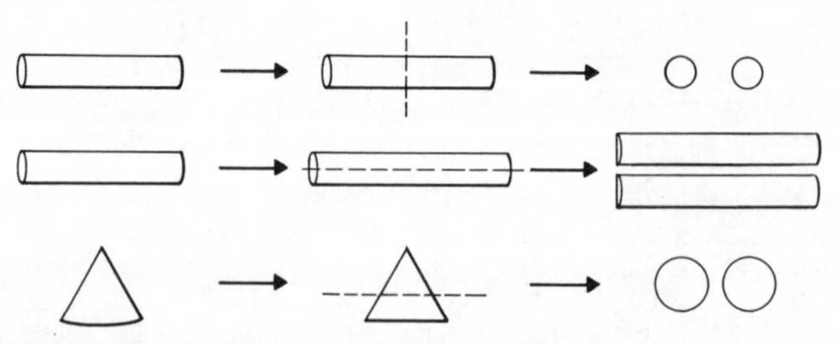

Figure 2–19. Objects cut along different planes.

figures. Beginning around the age of seven or eight years the child draws very rough unfolded figures and progresses until at the age of eleven or twelve years most children are able to correctly mentally unfold at least the cylinder and the cone. The cube and pyramid are more difficult and the ability to unfold these shapes is not achieved until a year or two later. A related study by Good (1973), however, suggests that even many adults have not developed the ability to mentally relate the rotation of two-dimensional figures to their three-dimensional solids. Three items on a 16-item instrument related to the subject's ability to mentally coordinate perspectives in going from two dimensions (a sketch of a simple figure) to three dimensions (the corresponding solid). Figure 2–20 shows the first item where *A* represents a piece of cardboard to which a string S is attached. Five possible responses that were available to choose from are also shown in the figure.

The specific question asked was, "If this piece of cardboard *A* was held by the string S and lowered into a vat of warm butter and turned by twisting S, what would the resulting 'hole' look like?" Of the 97 responses by the college seniors, 6 chose upright cone, 72 chose inverted cone, none chose double cone, 5 chose diamond shape, and 14 chose none of the above. Over one fourth (26 per cent) of the students were unable to correctly rotate the triangle in their mind to form the cone!

The second question was similar to the first except the figure used

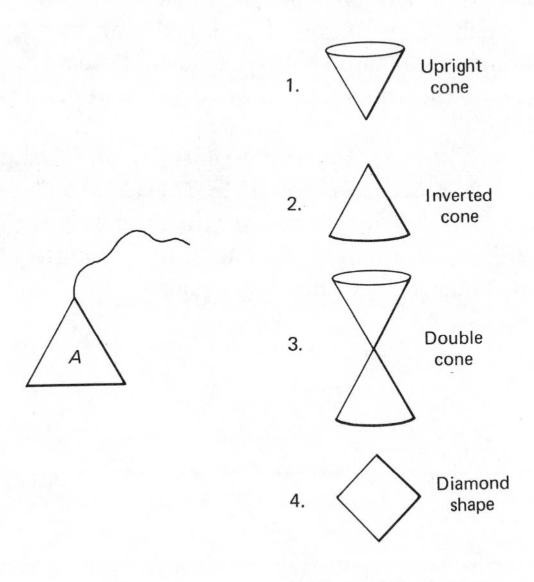

Figure 2–20. First question on rotation of a plane surface with possible responses.

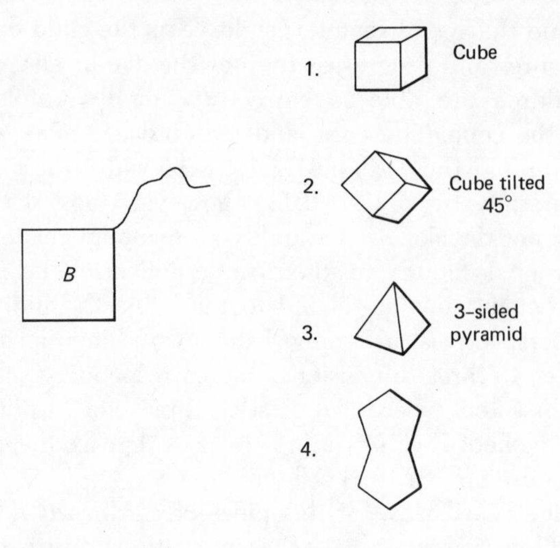

Figure 2–21. Second question on rotation of a plane surface with possible responses.

was a square instead of a triangle. Figure 2–21 shows the square and the five possible answers.

Of the 98 responses recorded on this second question, 6 chose answer 1, 28 chose answer 2, 15 chose answer 3, 6 chose answer 4, and 43 chose answer 5. Although 43 of these college seniors chose answer 5, none of the above, correctly, the remaining 55 or 56 per cent of the students were unable to do so!

These two items certainly raise doubts that there is any "stage of completion," as suggested by the data reported in *The Child's Conception of Space,* in a person's ability to imagine the rotation of a surface. The third item is represented in Figure 2–22 and strongly suggests that experience alone is not sufficient for the development of a fully coordinated perspective ability.

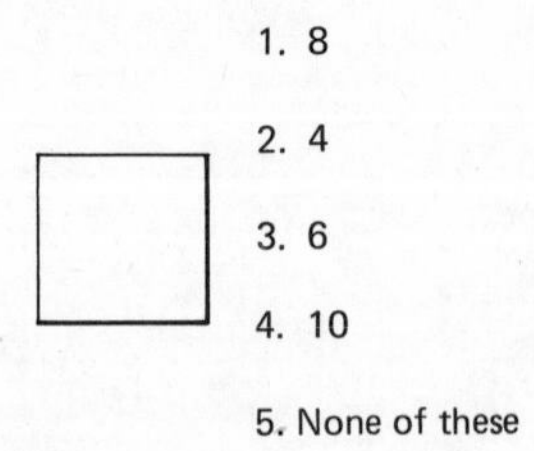

Figure 2–22. Question on edges of a cubic box.

This question asked students how many edges of the box were not seen from the front view. Although 60 correctly responded by circling the first choice, 8, 38 students or 39 per cent of the total were unable to make the correct choice. Twelve students chose 4 edges, 5 chose 6 edges, 7 chose 10 edges, and 14 chose none of the above. These results are rather surprising since all of the 98 students had certainly seen and handled dozens of boxes in their past experiences. Experience alone was certainly not sufficient to ensure the coordinated perspective required to imagine the box from various angles or to mentally unfold the box and count the folds.

More than anything else, the responses by these college seniors reinforce the position that a "final" stage is not reached in the development of perspective coordination. Further direct experience or lack of it will influence further development of the many and varied abilities related to spatial relations.

From Projective to Euclidean Space

The final part of *The Child's Conception of Space* by Piaget and Inhelder is devoted to a study of the child's transition from projective to Euclidean space. A brief summary of this transition period is presented here with a more detailed account of the development of length-area-volume relationships and coordinate systems to follow.

Transitional stages between projective and Euclidean space from a mathematical standpoint include conservation of parallelisms, angles, and distances. Piaget used various experimental situations with children to determine if the same transitional stages were evident in psychological or cognitive development.

A. Conservation of Parallelism

To determine the stages of development in the conservation of parallelism, a pair of "lazy tongs" shown in Figure 2–23 were used in interviews with children by Piaget and Inhelder.

Each child was asked to draw what he or she thought would happen to the tongs in a closed position (*D*) if they were opened by pressing the handles together. Not until after the age of seven or eight years could the children draw with any accuracy the transformations of the rhombus. After this point, there was an obvious attempt to keep the opposite sides of each rhombus parallel to one another. By the age of nine or ten years most children were able to maintain parallelism in the approximate changes in area as the tongs were extended.

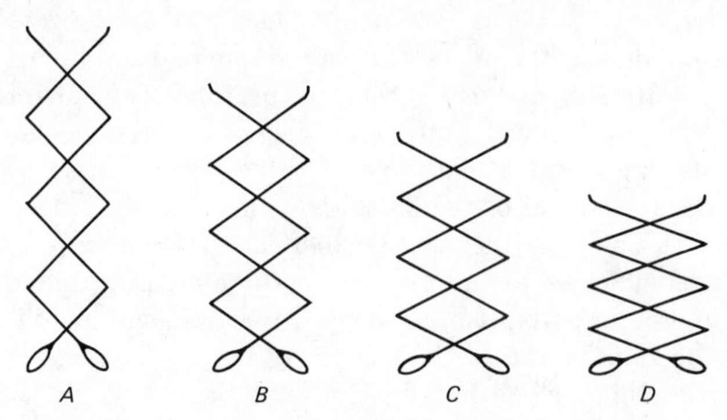

Figure 2–23. Lazy tongs used for rhombus transformations.

B. Conservation of Angles

The specific problem studied in this section is how proportions are perceived, represented, and extended to various shapes. We saw that parallelism was operationalized after the age of nine or ten years, which is one important aspect of proportions with squares, rectangles, and rhombuses. To discover how children discovered the similarity of triangles, a fairly lengthy set of procedures was developed. Without going into the details of the procedures, it was found that the attainment of true proportionality for both sides and angles of various triangles was not attained until the age of eleven or twelve years. Similar results were obtained when the shapes involved were rectangles. Proportionality with open figures also developed in a similar way, suggesting that the general concept of proportionality is fairly constant over a wide range of geometric figures. Other studies on the growth of children's ability to deal successfully with metric proportions, namely Lovell and Butterworth (1966) and Lunzer (1965), support the conclusions of Piaget and Inhelder but suggest that genuine understanding is not achieved until the age of fifteen years or even later, except for advanced students.

C. Vertical and Horizontal Coordinates

A brief account of the development of a coordinated reference system is presented here with more detail given in a following section. Locating objects in an exact way in two-dimensional space requires the coordination of vertical and horizontal axes. Number pairs are often used with children as early as third grade as a means for "basic" graphing techniques. As the data in this section show, such a task can only be memorization of a series of steps for most children.

Piaget has pointed out that the essential characteristic of a reference

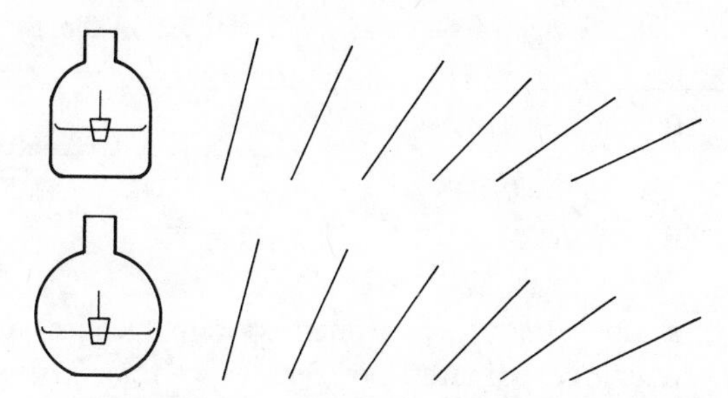

Figure 2–24. Bottles for studying reference system.

system resides in the possibility of coordinating positions and intervals without limit. This required more than a static reference system. A beginning step to the development of such a flexible reference system is the coordination of the vertical and horizontal in our physical environment. An ingenious method for studying the development of a vertical-horizontal reference system is found in *The Child's Conception of Space*. To study the child's conception of the horizontal coordinate in his physical environment, two narrow-necked bottles were used, one with straight, parallel sides and one with rounded sides (see Figure 2–24).

The basic idea is to ask the child to draw in the water level when the bottles are tilted at various angles. To represent the vertical axis, a cork is floated inside each of the bottles with a needle stuck into the cork and at an angle of 90 degrees with the water level. Drawings by the children indicated that it was not until the age of nine or ten years that they were able to coordinate correctly the vertical (needle) with the horizontal (water level). Before this time, the children draw the water level with reference to the bottle itself rather than to the surroundings.

A more recent study by Green et al. (1972) reported the types of water level drawings chosen by second-grade students (Figure 2–25). Table 2–8 shows the choices of the 24 second-grade students when shown the six drawings in Figure 2–25.

At the fifth-grade level, the same test revealed that 75 per cent of the 24 students interviewed chose *B* while the remaining 25 per cent chose *F*.

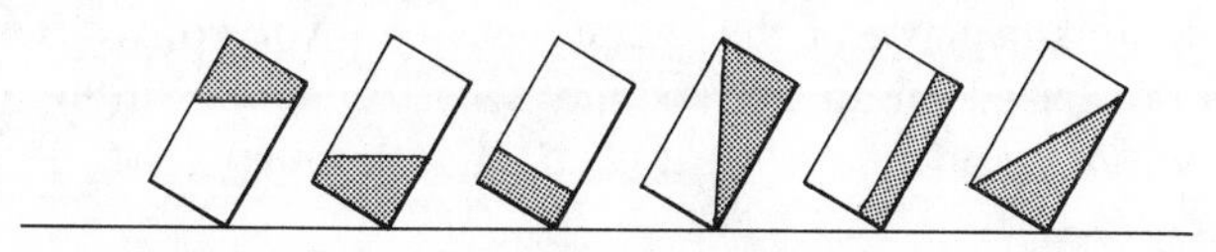

Figure 2–25. Water-level drawings.

TABLE 2–8. **Range of Choices on Water Level Test**

A	B	C	D	E	F
0	7 (30%)	2 (8%)	0	2 (8%)	13 (54%)

The development of a coordinated reference system occurs at approximately the same time as a coordinated perspective system. These developments mark the beginning of a comprehensive Euclidean and projective system, one that is able to cope with such concepts as movement and speed, as is seen later in Part Two. For now it is hoped that the reader has a "feeling" for the gradual development of space from topological through projective and Euclidean space. The more precise measurements of length, area, and volume relations as used in so many science concepts and principles are now studied in some detail.

Length and Distance Relations

In *The Child's Conception of Geometry* (1948b), Piaget, Inhelder, and Szeminska reported the results of experiments with children on length, area, and volume relations and the development of a coordinate system with which space can be quantified through exact measurements. Their book represents an extension of the slightly earlier book on *The Child's Conception of Space* but deals wholly with Euclidean concepts where angles, lengths, sides, distances, and so on are conserved. A square is no longer equivalent to a circle as in topology or a doughnut to a tire as in projective space. A square is not even equivalent to another square unless the sides are equal. The exactness required in the field of Euclidean geometry implies a much more quantitative approach.

Although we deal with length and distance in the same section, Piaget makes an important distinction between the two concepts, which is followed here. Distance is used to refer to the separation of objects, that is, to empty space. Length refers to the linear size of an object, that is, filled space. Locating a point in space involves distance measurements and it is the development of such measurement concepts that allows the transition from topological space to Euclidean space. Although it may seem that distance relations might be much more difficult than length relations because of the "empty" space involved, we see that both concepts are based on an understanding of order and interval between points or sites and not a great deal of difference exists in their development.

The method used to investigate distance relations was to set two trees, identical in height, about 50 centimeters apart on a table. A cardboard screen, slightly higher than the trees, is then placed between the objects and the child is asked if they (trees) are still as near or far apart. A movable window in the screen is then opened and the question is repeated. After pursuing reasons for the child's responses, a large cube, higher than the trees, is substituted for the screen and the process is repeated. By using these and other variations such as altering the height of one or both of the objects (for example, trees) the experimenter can assess the child's understanding of the constancy of the distance between the objects whether the space is empty or filled. Until the age of seven or eight years, children's responses show that they do not "conserve" the distance between the two objects. This is also about the time when the straight line begins to become operational and when the operations of ordering by length are understood.

Since empty space (distance) was not conserved by children until the age of seven or eight years, the next logical step was to determine whether filled space followed a similar pattern. Children are presented with a short straight stick and a longer "curvy" snake made of plasticene. The ends of the curvy object were made to coincide with those of the shorter stick (see Figure 2–26) and each child is asked whether the two sticks are the same length.

If the child's response is "the same," the experimenter makes up a story about two ants, each walking along one of the objects and asks if one would have to walk farther. Finally, the twisted snake is straightened out and then twisted back to its original shape and the questions are repeated. As with distance, it is about the age of seven years before the children give correct responses with logical reasons. Up to the age of five or six years, they concentrate on the end points and thus give incorrect responses.

A more recent replication study by Lovell et al. (1962) involved 70 children from the ages of five to nine years using similar materials and interview procedures for both distance and length relations. As can be seen by comparing Tables 2–9 and 2–10, conservation of length develops somewhat earlier than similar distance relations. These results are not consistent with those reported by Piaget and raise the question as to why there is such variability on the distance relations results. Piaget has

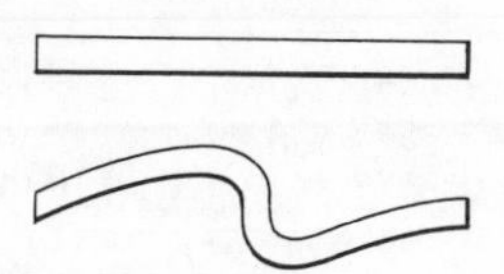

Figure 2–26. Conservation of length materials.

TABLE 2–9. Distance Relations

Age	Nonconserver Stage 1	Transitional Stage 2	Conserver Stage 3
5	9 (90%)	1 (10%)	0
6	5 (33%)	10 (67%)	0
7	8 (53%)	4 (26%)	3 (21%)
8	4 (26%)	7 (47%)	4 (26%)
9	1 (7%)	10 (67%)	4 (26%)

theorized that both concepts (distance and length) are based on an understanding of order and intervals between sites, which suggest similar results in interviews. One can only speculate in the absence of convincing data on reasons for these discrepancies. One possible explanation is that although both relations have similar conceptual bases, distance is a slightly more abstract notion simply because the space is, in fact, empty. With length relations, objects can be seen, touched, and manipulated so that sizes and lengths can be assimilated directly into the child's experience. Another possible explanation for the discrepancies between results in the two studies is that the nature of the interview situations was somehow different. DeLacy (1967) has reported that conservation of length results can be completely masked and confused if optical illusions are introduced (see Figure 2–27). Table 2–11 shows the results of interviews with 140 children, and it is easily seen that even by eleven years only 34 per cent are judged conservers of length, which is far less than "typical" results for conservation of length experiments. A similar study was done by Allen (1972) with fourth and fifth grade children with fairly similar results.

Though these two studies deal with length rather than distance, where the problem seems to lie, they do suggest that variations can result from using different materials in interview situations, which has

TABLE 2–10. Length Relations

Age	Nonconserver Stage 1	Transitional Stage 2	Conserver Stage 3
5	3 (30%)	3 (30%)	4 (40%)
6	2 (14%)	5 (33%)	8 (53%)
7	1 (7%)	4 (26%)	10 (67%)
8	0	4 (26%)	11 (74%)
9	1 (7%)	4 (26%)	10 (67%)

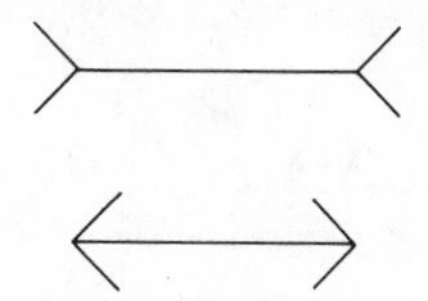

Figure 2–27. Mueller–Lyer arrowheads length illusions.

been confirmed in studies not dealing with length or distance. If nothing else is learned from this detour, it once again points up the many variables involved in "knowing" something about the cognitive development of children. Variations in the experimental data, though not necessarily requiring substantial changes in Piaget's theory of development, should cause every serious student of the theory to question and reconsider ideas from all possible viewpoints.

Another experiment by Piaget et al. confirmed their previous findings on length conservation. The basic procedure is to take two objects of equal length and then cut or break one of them a few times and arrange the pieces end-to-end in a nonlinear fashion (Figure 2–28). After the child has agreed initially that both objects are of equal length, the broken pieces are laid in an irregular, or at least nonlinear, pattern and the child is again questioned about their lengths. The results were similar to those found with the other conservation of length experiments. As an extension of this length experiment, children were asked to actually measure the two objects after one of them had been subdivided and glued to a piece of cardboard. The measuring devices consisted of cardboard strips equal to or longer than the shortest of the glued pieces on the cardboard. Up to about the age of six years, the children are, of course, nonconservers of length and their attempts at measuring

TABLE 2–11. **Effect of Optical Illusion on Length Conservation**

Age	Per Cent Able to Conserve Length
6	11
7	17
8	39
9	37
10	10
11	34
12	63

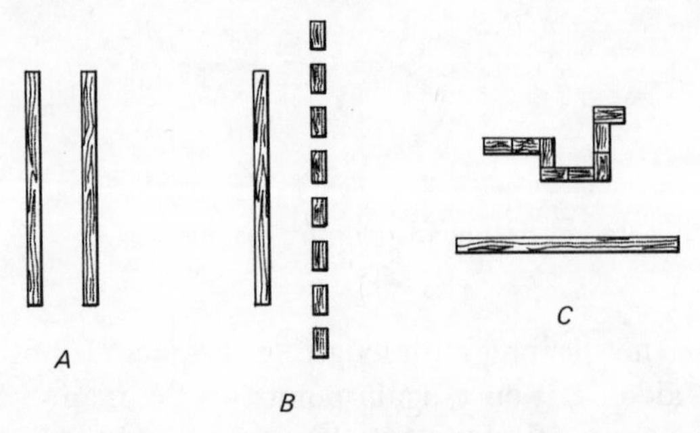

Figure 2–28. Conservation of length after subdivision.

the solid and the broken objects glued on the cardboard are not systematic or accurate. At the age of six or seven years, the children begin to show the understanding of the measuring device as a "unit" that can be used over and over in measuring the length of the object. This, combined with the realization that the object can be measured beginning from either end (object length is permanent), allows the child to achieve fairly accurate measurement by about the age of eight or nine years, although some children will take longer.

To see just how accurately the children could subdivide a straight line, a series of pairs of strings, each with a bead, was presented to each child. Figure 2–29 shows the various arrangements of the pairs of strings.

Various unmarked objects were also available for the child to use as he or she wished in response to the interviewer's questions (stick, paper strips, pencil, and the like). Questions such as the following were asked:

1. I'm going to go this far with my train; where should yours go if you want to travel just as far?
2. If I start from this end (opposite to child's bead) and travel to here, where should your's (bead) go to travel as far?
3. How can you be sure they traveled the same distance?
4. If we both travel together and then stop at the same time, will we travel the same distance?

From the age of four to six years, the children can deal only with pair *A* where the lengths are equal and they are aligned and parallel. The children in this age group are unable to compare the distances traveled when the starting points are staggered. By the age of seven or eight years they can measure using an unmarked unit measure on strings that are staggered but equal. During this second stage of development in subdividing a straight line, however, the children are still unable to

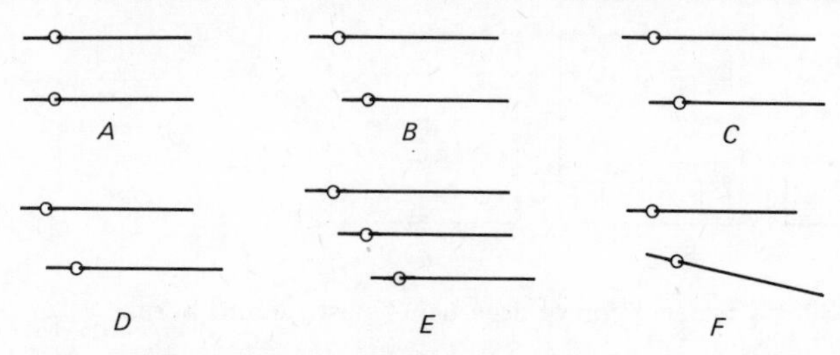

Figure 2–29. Strings with beads used for subdivision of a line.

deal with unequal and nonparallel lines, indicating that they are not yet fully operational in subdividing the line. By the age of eight or nine years, the third stage appears where all pairs of lines can be subdivided correctly by transferring unit measures from one line to another even though the lines may be unequal, nonaligned, or nonparallel. In comparing these results with those reported by Lovell et al. (1962) in which the experiments were replicated, we see discrepancies again. Lovell reported that even by age nine only 26 per cent of his subjects were able to exhibit an operational understanding of the subdivision of a straight line. As with length and distance conservation, Lovell's results suggest that many children do not achieve an operational stage with some concepts nearly as soon as implied by Piaget and his co-workers.

Area

The concept of area is important in science in both a qualitative and a quantitative sense. Comparing the surface area of objects in a qualitative sense involves estimations of size and shape, whereas the exact measurement of each area requires the conceptual development of a coordinate system. Just as a line may be defined as a continuous set of points in one dimension, an area or surface may be defined as a continuous set of points in two dimensions. Before an actual measurement of some area can make sense to a child, the area must be conserved in the child's mind even though a change in shape occurs. A rectangular piece of paper retains the same amount of area no matter how it may be cut and fitted together again. If it is cut in half lengthwise and placed end-to-end, the reduction in width is exactly compensated by the gain in length. Though the compensation principle or concept of conservation may be completely evident to adults, research has shown that such is not the case with young children. Many different interview tasks have been developed and patterned after those originated by Piaget et al. (1948b), but the basic idea is the same for all versions. The child is shown two pieces of paper, cardboard, or the like,

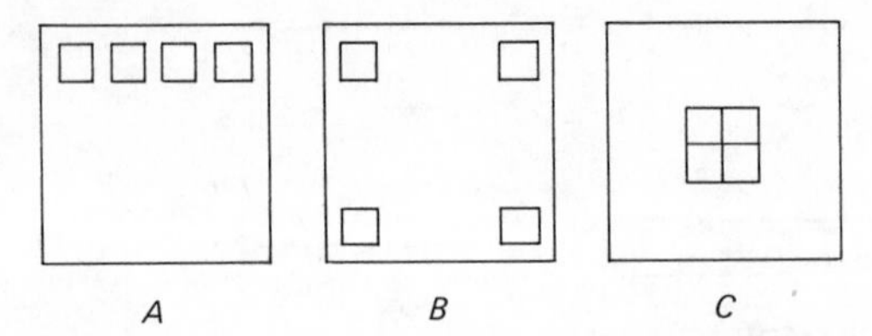

Figure 2–30. Conservation of area using pasture and barns.

which are identical. One of these pieces is then changed in terms of shape and the areas are again compared. One study by Good et al. (1973) used the well-known "pasture and barns" technique shown in Figure 2–30.

The child is shown a green square of cardboard 8 or 9 inches on a side with four "barns" approximately 1 inch square. The barns are arranged in a certain pattern, such as in *A*, and the child is told that cows have all of the green grass available to eat. Then the barns are changed to a different arrangement, as in *B* or *C*, and the child is asked if the cows would now have more, less, or the same amount of grass to eat, and why? Two-hundred forty-nine children in grades 1–5 were individually interviewed near the beginning of the school year and the results of the interviews are shown in Table 2–12.

These results are not in agreement with those reported by Piaget, et al. (1948b). Their data suggested that by the age of seven or eight years, children could conserve the exposed area of the pasture no matter how the barns were arranged. Table 2–12 data show that not even by fifth grade or about the age of eleven years do more than 58 per cent of the children conserve the area represented by the exposed "pasture." At the second grade level, where most children are seven or eight years of age, only 17 per cent were able to overcome their perceptions to conserve area. Although the nature of the interview was somewhat different in each study, it is difficult to explain the large discrepancies in terms of these differences alone. No attempt is made at this point to

TABLE **2–12. Results for Conservation of Area Task**

Grade	Category 1 (Conserver)	Category 2 (Transitional)	Category 3 (Nonconserver)
1	5 (10 per cent)	5 (10 per cent)	39 (80 per cent)
2	8 (17 per cent)	6 (13 per cent)	34 (70 per cent)
3	20 (40 per cent)	3 (6 per cent)	27 (54 per cent)
4	26 (52 per cent)	4 (8 per cent)	21 (40 per cent)
5	30 (58 per cent)	6 (12 per cent)	15 (30 per cent)

try to explain the differences in the results, but the reader should be aware of variations, sometimes quite large, in what might seem to be similar studies. It certainly suggests that caution should be exercised in generalizing the results from a single study to all situations. This uncertainty combined with the range of developmental abilities within a single grade level, usually at least two to three years, should caution the classroom teacher and curriculum developer against assuming that all or even most children will have attained certain cognitive abilities by a given age. As Table 2–12 data show, there is a wide range of abilities on this specific task at each grade level. For the teacher of secondary science students who might believe that research on something like area conservation has no relevance to his/her learning problems, a recent study with college students, Good (1973), suggests otherwise. Eighty-two college seniors were given a paper and pencil test on area conservation and over one third failed to choose the correct figure that represented a conserving response! Although the limitations of group-administered paper and pencil tests are acknowledged, sufficient doubt is raised about conceptual abilities even beyond the eleven to twelve year stage that Piaget has identified as the onset of formal or abstract thinking.

Although the "cows and barns" interview was used by Piaget to investigate a child's ability to deal with congruent areas, by the very nature of the questions and materials the child's ability to "conserve" areas is also determined. Another interview procedure was developed to test specifically for area conservation using 12 identical cardboard squares. Without going into the details of the interview procedure, the squares were arranged in different ways and the child was asked if the shapes were the same size, that is, if one was larger than another. An alternative but similar method is shown in Figure 2–31. Rectangle *A* is cut into two right triangles and placed together as in *C* to form a triangle.

The child is then asked to compare shapes *A* and *C*. Results reported by Piaget et al. (1948b) for this interview were similar to those reported for the "cows and barns" interview with the child accomplishing conservation after the age of seven or eight years. Keep in mind that this conservation requires a kind of reciprocal compensation between the

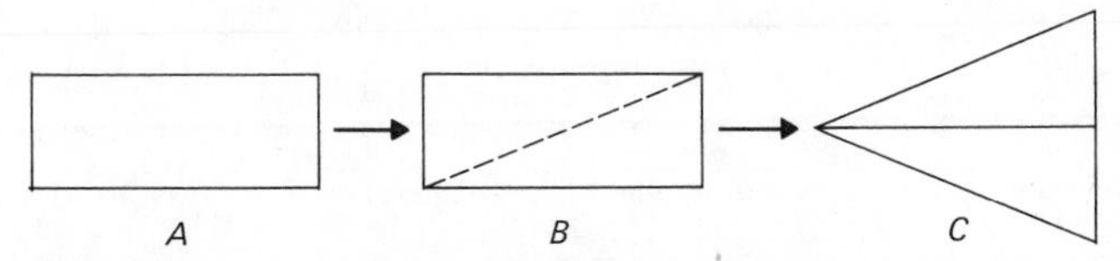

Figure 2–31. Area conservation with different shapes.

vacated and the occupied sites as with the barns or the squares. In fact, all of the conservation tasks are so named because of the requirement that one must be able to conserve some original quantity or state. To be fully operational, both reversibility and logical necessity are required to be a conserver of something, but as we have seen and shall see in more detail later, there is no one age at which a child is a conserver of all things. Reversibility may be exhibited by a child in a number conservation task but not in area conservation. Likewise, logical necessity (nothing has been added or taken away) may be used by a child of nine years of age during an area conservation task but not for a weight conservation task. Uncertainty and variability among children's responses to various interview measures is a somewhat disconcerting but valuable piece of information for both the teacher and curriculum developer. A knowledge of the lack of uniformity of responses to various cognitive measures allows one to take appropriate steps in instructional design, which is done later in this book.

A study reported by Lovell (1971) and conducted by M. Caltieri tested children's ability to measure area by a form of superposition using small right triangles as the area measures (see Figure 2–32).

Eight of the small triangles were supplied along with the larger square and rectangle, and the children were asked to compare the area of the square and rectangle. Various questions were asked to assess each child's ability to measure the areas including, "If both were fields, which would have more grass?; How many tiles (small triangles) are needed to cover the square or rectangle?; How can you tell exactly half of the square (or rectangle) has been covered by the eight triangles? Table 2–13 shows the results of this study with 90 children, ages five to ten years. The stages of development are identified by I through III *B* and follow the typical "stage designations" used by Piaget and his co-workers in Geneva. Stage I typically means that children were unable to understand much of what was to be accomplished, Stages II *A* and II *B* represent some understanding but errors are still made or reasons are not logical, and Stages III *A* and III *B* include children who successfully complete the task (*A*) and finally are able to generalize to similar situations with accuracy and understanding (*B*). Table 2–13 indicates

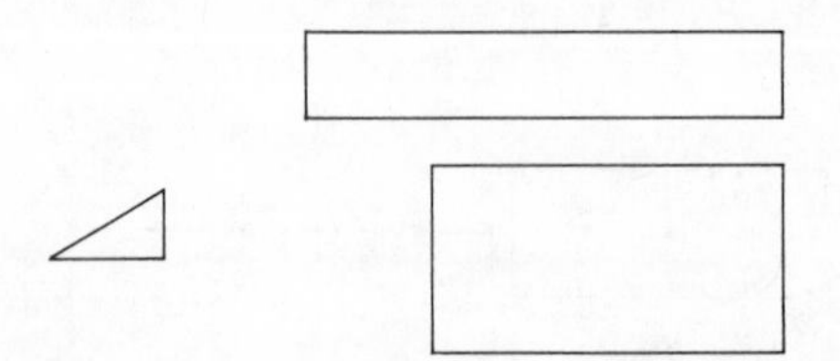

Figure 2–32. Area measurement shapes.

TABLE 2–13. Stages of Development in Area Measurement

	Age					
Stage	5	6	7	8	9	10
I	1	0	1	0	0	0
II *A*	11	7	4	3	0	0
II *B*	0	7	8	3	1	0
III *A*	3	1	2	2	3	0
III *B*	0	0	0	7	11	15

that before the ages of nine to ten years most children are unable to measure regular areas such as a square or rectangle using unit areas when superposition is required (too few unit areas provided to measure the larger area).

These results reported by Lovell are, to some extent, in disagreement with those reported by Piaget et al. (1948b) although the tasks are dissimilar in two important aspects. Namely, the smaller units in the Piaget study were supplied in sufficient quantity to cover the larger shape to be measured and there were three different smaller shapes: squares, rectangles, twice the size of the squares, and triangles one half the size of the squares. The concept of transitivity (if $A > B$ and $B > C$, then $A > C$) is used by Piaget to explain the development of the ability to measure area by superposition. Stage III or operational understanding is achieved after the age of seven to eight years as demonstrated by the child's ability to measure by superposition, although as we saw in the results reported by Lovell, another type of superposition is not achieved until after the age of nine or ten years.

Unit iteration was also investigated by Piaget whereby a child's ability to measure an area, say $9A$, using one unit measure such as A or $\frac{1}{2}A$, was investigated. This process of using one unit measure over and over again to determine the total area of a larger shape is known as iteration and is obviously very basic to any sort of quantitative measurement in length, area, volume, weight, mass, and the like. The fact that Piaget reports that children accomplish unit iteration of area after the age of seven or eight years is somewhat out of line with the Lovell study and with a replication study by Lovell, Healey, and Rowland (1962). Interviews with 70 children from five to nine years of age were administered individually using the same materials and procedures as reported by Piaget, but results showed that most children attained Stage III *B*, operational understanding, after the age of nine years.

A more recent study by Lunzer (1968) reported interesting results

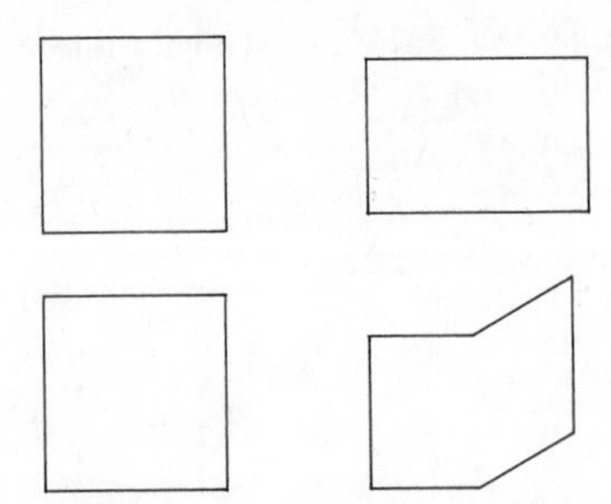

Figure 2–33. Perimeter–area relationships with constant perimeter.

with a test designed to relate length, or more specifically, perimeter to area. Figure 2–33 shows the shapes of string on a pegboard presented to the child who was then asked to compare areas and perimeters. One hundred and forty children, ages five to fifteen from two different locations, Geneva and Manchester, England, were individually interviewed using both transformations, $A \rightarrow B$ and $C \rightarrow D$.

If one studies each of the transformations, it can be seen that whereas perimeter is conserved, area is not conserved. Results of the 140 interviews on each task showed that children at the stage of concrete-operational thought (ages seven to eight to eleven to twelve years) conserved *both* area and perimeter, whereas children at the stage of formal operations (logical thought) were able to dissociate the two quantities and analyze them separately. This, of course, suggests that the perimeter-area task actually tests formal operational thinking rather than merely combining mental operations developed at the point of length and area conservation. The essence of Lunzer's experiment was repeated in a study by Good and Shymansky (1973) with 242 children in grades 1–5. A loop of yarn was shaped into a square, approximately 8 inches on a side, by using four pins stuck into a piece of soft ceiling tile. The pins were then removed and relocated to form a rectangle approximately 10 x 6 inches. A story about fencing and enclosed grass accompanied each figure, and the child was asked if there would be more grass to cut in one of the shapes as compared to the other and why. Then, the interviewer asked each child to make a fence with the yarn loop so that the least amount of grass would be enclosed and finally to make one that would enclose the greatest amount. One child in fifth grade, one in fourth, and two in third were able to respond correctly to all three parts of the interview, although the two third-grade children were uncertain in their responses. The remaining 238 children in the sample responded incorrectly to at least one of the three questions. First- and second-grade children were completely unable to comprehend the concept of limits even though a relatively large number stated correctly that the area of the rectangle would be less. Almost invariably, the reason given by the younger children (six to eight years of age) for the smaller area of the rectangle was that "it looked smaller" or "the square is just bigger." Table 2–14 shows the results of the square-to-rectangle question and the related lower and upper limits problems.

These results would appear to confirm Lunzer's hypothesis that perimeter-area relationships involving nonconservation of one of the quantities, in this case area since the same piece of yarn is used in all cases, require formal operational thinking. Only two of the 242 children, or less than 1 per cent, were able to correctly state that the rectangle enclosed less grass and could then close the loop for the lower limit and form a circle for the upper limit. A very interesting trend in the results can be seen in Table 2–14 by looking at the first column of data. Far more children in grades 1 and 2 gave the correct answer, that the rectangle enclosed less area than the square, than did children in grades 4 and 5. Approximately four times as many younger children (aged six to eight years) gave correct responses as did the older children (aged nine to eleven years). The apparent paradox is solved, however, if one remembers that very few children in grades 1 and 2 are conservers of area (see Table 2–12). Because of this, they rely primarily on their perceptions and the square is seen as "fatter" and thus "bigger." As the children get older, more and more develop the ability to conserve area and are "tied" to the conserving response, unable to separate the perimeter conservation (yarn loop) from area nonconservation.

Since the hypothesized ages for the beginning of formal or abstract thinking are about twelve to thirteen years of age, one would expect college seniors to give correct responses to the perimeter-area task. A recent study by Good (1973) fails to confirm this expectation, however. Two thirds of the 97 students thought that the area would be the same, and only 10 per cent said that the square would enclose greater area. These results strongly suggest that there is no assurance that adults

TABLE 2–14. Perimeter-Area Relationships

Grade Level	Rectangle has less area.	Rectangle and square same area.	Rectangle has more area.	Lower limit OK.	Upper limit OK.	Both limits OK.
1	17	17	12	0	0	0
2	16	23	9	1		
3	8	30	9	4	2	2
4	3	43	3	6	5	5*
5	5	41	4	22	13	9*

* *Only one student also said the rectangle has less area (correct answer).*

reach formal operations, as predicted by Piaget's theory. More is said of this "extension" of concrete operations into adulthood later, but it should be kept in mind that the very real possibility exists that many, perhaps most adults operate at a concrete level. By definition, this means that they are unable to go beyond the real, concrete world into what is hypothetically possible. In the perimeter-area task, the upper and lower limits correspond to the hypothetical situation, although the fact that the loop of yarn is present also allows for a concrete approach.

VOLUME

Volume would appear to be closely related to quantity or amount of substance, but as we see they are distinctly different concepts, at least from a developmental standpoint. Lovell and Ogilvie (1961) reported the results of a study with 191 British children, ages eight to eleven years. Each child was interviewed individually on internal volume, occupied volume, and displacement volume.

Internal Volume

Twelve cubes were stacked as a 2 x 2 x 3 (last numeral is height) and another 12 cubes were in a 2 x 3 x 2 arrangement. The child was told that each structure represented a box for holding bricks and was then asked if there would be as much room for bricks in one box as in the other (see Figure 2–34).

After recording the child's response and reason, arrangement *A* was changed to 1 x 2 x 6 (*C*) and then compared with arrangement *B*. Many children found the first question comparing arrangements *A* with *B* to be easier than the second comparing arrangement *C* with that of *B*. Eighty-six per cent of the eight to nine-year group gave correct replies for the *A* versus *B* comparison, whereas only 66 per cent of that group correctly responded to the *C* versus *B* structures. This suggests that "conserving" responses are affected by the nature of the block arrangements, at least until the age of ten or eleven years where the results were not very different.

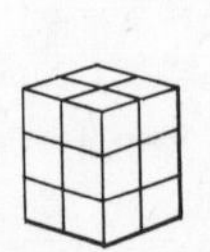

Figure 2–34. Cube arrangements for internal volume task.

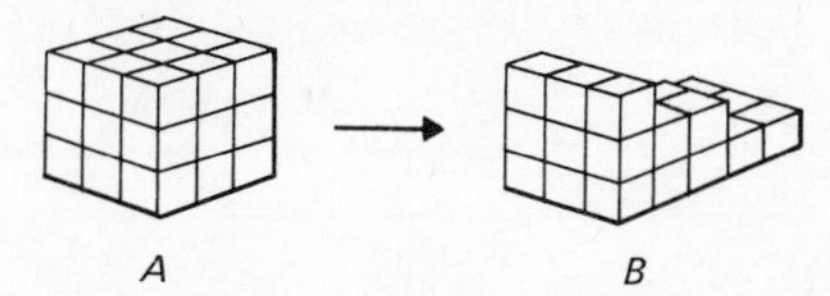

Figure 2–35. Internal volume with 27 cubes.

A more recent study by Good et al. (1973) lends strong support that the nature of the block arrangement affects the likelihood of a conserving response. The interviewers in this study used 27 cubes in a 3 x 3 x 3 arrangement and then changed it as shown in Figure 2–35.

A story accompanied the task about a tiny bird inside the "hollow" bird cage having flying room in both structures *A* and *B*. At the age of ten or eleven years, at which point Lovell and Ogilvie reported that almost 100 per cent of the subjects were conservers of internal volume, Good reported that only about 50 per cent of the 172 subjects could be categorized as conservers. The rather substantial differences could, of course, be attributed to other procedural differences, but the most obvious difference was the number and arrangements of each block structure. A further study by Good (1973) with 97 college seniors also raises questions about the extent to which even adults are able to conserve internal volume in the face of spatial rearrangements. Only 60 per cent of the seniors were able to choose the conserving response (same flying room) from among five possible choices. Although the task was administered with sketches of the block structures rather than the actual cubes, Phillips (1971) found no significant differences in results between presentations with the blocks and those with only drawings of the blocks.

In *The Child's Conception of Geometry* (1948b) Piaget et al. reported using 36 cubes 1 centimeter on a side and constructed "houses" of various shapes and asked children if one had more room than another. They found that by the age of eight or nine years most children could conserve internal or "interior" volume, but exact measurements in three dimensions are not achieved until later.

Displacement Volume

Piaget used a similar task to determine the child's ability to conceptualize occupied or displacement volume. Thirty-six small metal cubes were placed at the bottom of a bowl of water and the child was asked if the water level would change if other constructions were made, that is, 2 x 1 x 18 or 2 x 2 x 9. By the age of eleven or twelve years, Piaget found that children were capable of what he termed "true" conservation where surrounding space is related to the occupied volume. They dis-

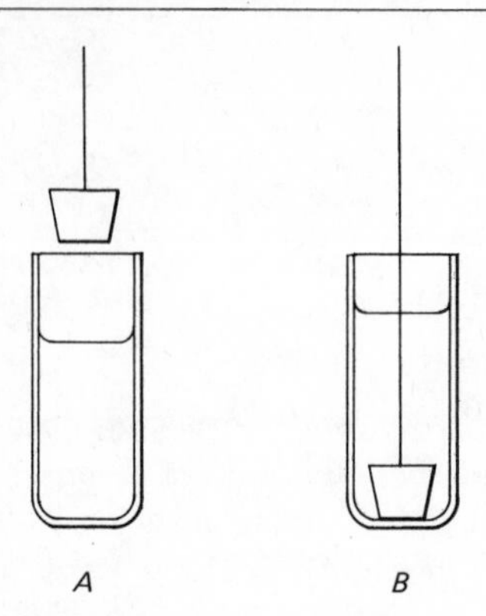

Figure 2–36. Displacement volume materials.

cover that not only is the interior invariant but the surrounding space is conserved, as well.

The study by Lovell and Ogilvie (1961) described previously also investigated displacement or occupied volume and their results are essentially in agreement with those of Piaget. They did find, however, that not until about eleven years of age do 50 per cent of the children realize that the amount of water displaced by an object is independent of the size of the container. Similar results were obtained when children were asked about the displacement potential of two bricks of the same size and shape but of different weights.

Figure 2–36 shows the materials used by Good et al. (1973) to assess the children's development of displacement volume. Each child of the 178 interviewed was asked what would happen to the water level in *A* if the stopper were to be lowered to the bottom. Then the interviewer lowered the stopper as in *B* and asked the child to predict what would happen if the stopper was pulled up to about the halfway point in the vial. Table 2–15 shows the results of the interviews with children in grades 1–5. It is not until grade 5 (ten to eleven years of age) that any

TABLE 2–15. Displacement Volume Data

Grade	Category 1 (Conserver)	Category 2 (Transitional)	Category 3 (Nonconserver)
1	0	5 (14%)	32 (86%)
2	2 (5%)	6 (15%)	33 (80%)
3	2 (5%)	11 (29%)	25 (66%)
4	3 (6%)	25 (50%)	22 (35%)
5	8 (25%)	13 (40%)	11 (35%)

appreciable number of children are able to respond correctly with a logical reason and even then only one quarter of the children tested at that grade level were conservers. The probability for a large increase of conservers in grade 6 (eleven to twelve years of age) would appear unlikely, particularly in light of data collected in another recent study, Good (1973), with college seniors. Two drawings, shown in Figure 2–37, were presented to the students with accompanying questions directed at assessing their ability to conserve occupied or displacement volume. In *A* the student was told (in writing) that the stack of cubes would be knocked over, and in *B* the steel cube is suspended by a pair of magnets such that the cube can be moved to any position within the beaker. Five choices regarding water level accompanied each question. Eighty-three per cent responded correctly to *A* and 71 per cent conserved the water level in *B*. The 17 per cent and 29 per cent of the students who were unable to conserve displacement volume again raise the question of when people become conservers of a particular quantity or state. The fact that an individual is eleven or twelve years or even seventeen or eighteen years of age apparently does not ensure operational understanding of concepts and principles related to science and/or mathematics learning. This is further substantiated in another study with college students by Towler and Wheatley (1971). They reported that only 44 of 71 subjects (70 per cent) gave conserving responses on a conservation of volume task.

Volume measurement in an operational sense requires a coordinated reference system that allows the individual to locate and relate points in space. The development of such a reference system is a lengthy, complex process and quite obviously involves concepts of length-area-volume relationships that we have just considered, primarily from the point of view

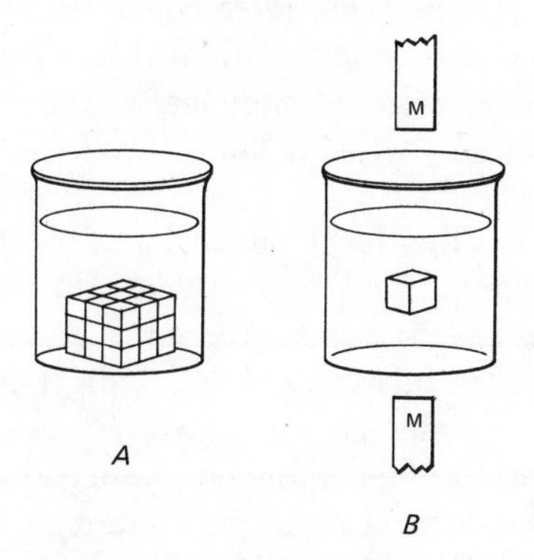

Figure 2–37. Drawings used for displacement volume task with college seniors.

of conservation. We now look at the development of abilities such as the location of points in space, angular measurement, and curves in space.

Coordinate Systems

Our study of length-area-volume relations thus far has emphasized conservation aspects rather than the development of a coordinate system used for exact measurements. Science usually attempts to answer questions such as "how" and "how much" and quite naturally requires means for attaining the answers. Physics, in particular, is concerned with locating bodies in space at any given time. Both Newtonian mechanics and Einsteinian (quantum) mechanics require that the experimenter be able to create any number of coordinate systems that can be related one to the other, in order to make meaningful motion measurements.

Piaget devoted the final part of *The Child's Conception of Space* and Chapters 7–10 of *The Child's Conception of Geometry* to analysis of the development and use of two- and three-dimensional coordinate systems. Plane and solid geometry deal directly with measurement in two and three dimensions, respectively, and thus assume the ability to visualize and represent such relations. Just as the child who is a nonconserver of weight or volume is severely limited in conceptualizing density, so is the individual who has not yet developed an operational coordinate system for measurement purposes. Most of the material presented in this section is found in Piaget's two books on space and geometry identified earlier.

Coordinates of Euclidean space provide a network of linkages among all points within a specified boundary, in one, two, or three dimensions. The construction of straight lines, parallels, angles, and the like mark the beginning of an operable reference system. This reference system develops first in only one dimension, then two, and finally in three-dimensional space. The reference system becomes more functional as the child develops the ability to coordinate positions without limit, by expanding the basic system.

An important initial ability involves the development and coordination of a stable two-dimensional coordinate system. The vertical coordinate is a rather "natural" one, just as the horizontal reference frame is commonly confronted by the child. Trees, telephone poles, walls, falling objects, lamps, table legs, fenceposts, and the like all suggest a "natural" position as a vertical reference system. Ceilings, tabletops, water, floors, level ground, and the like all suggest a natural horizontal reference system. We saw earlier that Piaget used a bottle of water with a vertical needle stuck into a cork to determine the child's ability to coordinate the vertical and horizontal, and it was not until the age of nine or ten years that operational coordination was achieved. In addition to investigating these so-called "natural" coordinates, Piaget and Inhelder (1948a) reported results on the development of generalized reference systems. They used heavy, square beads in a certain pattern and then asked each child to reproduce the figure using another set of beads. Rectangular strips of paper were also supplied and could be used as reference lines, if so desired. With these and other materials, also designed to assess a child's understanding of reference systems, it was found that children younger than six or seven years make no use of the reference strips, relying instead on perceptual characteristics. Until about the age of eleven or twelve years, most children can make qualitative but not quantitative applications of the reference systems. After the age of eleven or twelve years, the children coordinate position with distance using the paper strips in various arrangements as reference systems. An interesting finding confirmed by the work of Piaget and others is that young children can estimate the length of differently tilted lines more accurately than older children because they ignore orientation.

In *The Child's Conception of Geometry* a procedure for studying how children measure angles is described. Each child was shown a drawing as in Figure 2–38 and was then asked to make another drawing. The model drawing was placed behind the child so that he/she could not look at it while actually drawing, but could refer to it at any time. Rulers, paper strips, string, triangles, and compasses were available to each child, but up to about the age of six years no attempts at measurement were made. At the age of seven or eight years, the lengths of the lines *AB, DC* are measured but there are as yet no attempts to measure the angles *C D B* or *C D A*. By the age of nine or ten years, Piaget found that children could superimpose triangles to compare the angles but were still unable to use a linear measure (for example paper strip, ruler) to determine the line separation. Accurate angle measurement was not achieved until the age of ten or eleven years, the same age at which reproduction of a triangle is also achieved.

The achievement of angle measurement combined with the previously developed ability of distance measurement using unit iteration (eight or nine years of age) marks the operational level of a two-dimensional reference system. Mapping is an activity that utilizes an operational reference system and it can be seen in the actions of children

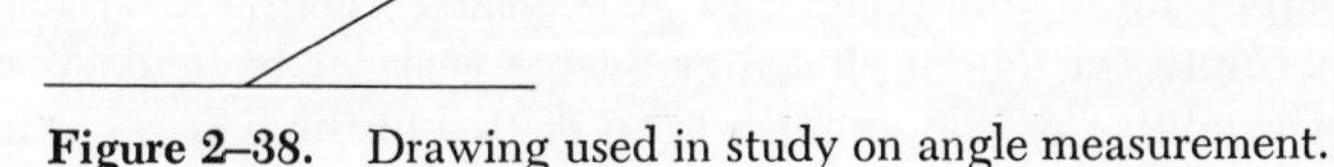

Figure 2–38. Drawing used in study on angle measurement.

that not until the age of eleven or twelve years can they make the necessary measurements. Piaget (1948a) has outlined eight sublogical operations that constitute an operational coordinate system:

1. Addition and subtraction elements.
   Operations of this type define relationships between points of a given figure.
2. Placement and displacement of objects.
   The order or placement of objects relative to each other are involved in this operation. A displacement of the objects entails an inversion of one or more of the elements of the series.
3. Reciprocity of references.
   This operation allows the individual to begin at the same reference point as the beginning of two or more separate (reciprocal) figures.
4. Inclusion of intervals or distances.
   The conservation of distance between two points is assured no matter what their orientation.
5. One-to-one multiplication of elements.
   An interval (series of points) multiplied by another interval yields a surface. That surface then multiplied by another series of points yields a volume.
6. One-to-one multiplication of placement and displacement relations.
   This operation defines the development of a coordinate system where intervals along one axis coordinated with intervals along the other result in a matrix or plane of points.
7. One-to-many multiplication of elements.
   This operation is one step beyond the preceding one and results in the measurement of figures where sides are not parallel to the matrix system, that is, triangles for two dimensions and tetrahedrons for three dimensions.
8. One-to-many multiplication of relations.
   This final operation is similar to the preceding one except that relations rather than elements are involved. It will be recalled from the sections on classification and seriation that elements define classes whereas comparisons between or among elements define relations. It might be useful to think of relations as dynamic in nature whereas elements are static. A part-whole relation defines one set of elements in terms of another larger set.

The development of these eight operations signifies the key steps required before an operational and generalized reference system is achieved, not until the age of eleven or twelve years or beyond. As with all stages of cognitive development, a span of two to three years or more can be expected in any grade level in terms of abilities among various children to construct and use a generalized coordinate system in two or three dimensions. Support for this variability has been seen in the many examples already reviewed as well as in a study by Robinson (1970) that is specific to coordinate systems. One-hundred twenty children in grades 1–6 were interviewed and it was found that wide variations existed in the ability of the children to use rectangular coordinates. Among the findings were these:

1. Children in the top 30 per cent in terms of apparent intellectual ability and achievement were competent in using rectangular coordinates.
2. Children in the mid-40 per cent were competent by the sixth grade.
3. Children in the lower 30 per cent developed the abilities required later than the sixth grade.

Without concerning ourselves with the nature of the interview tasks, criteria for determining competency, and so on, it is seen that wide variability exists in children's ability to use a fairly simple type of two-dimensional coordinate system.

Time and Motion

In Part One, time experiments were described briefly as examples of interviews with the use of manipulative aids. The treatment was brief and is extended here and combined with a study of the development of motion concepts.

The work that led to the eventual publishing of two books by Piaget on time (1946a) and motion (1946b) was suggested to him by Albert Einstein more than 15 years prior to the appearance of the two books. Einstein was apparently interested in the intuitive grasp of time and motion and the extent to which they are developmental in nature. All of us, regardless of formal learning in the areas of time and motion, seem to have an intuitive notion of such concepts as older, younger, faster, slower, before, after, and the like, as represented in events occurring around us daily. And as with almost all developmental abilities, we cannot remember with any accuracy what our "mental life" was like before we developed our current mental abilities. We encounter situations related to time and/or motion constantly even though they may not require a thorough understanding of Newtonian or Einsteinian mechanics.

Although we study the development of time concepts first, by the very nature of the general concept (try defining time), motion studies become an integral part of the process. Piaget referred to space as a "still of time" whereas time is "space in motion." Time can be ignored in geometrical studies but not when movement or motion is involved. A number of experiments were used by Piaget to test for the child's concepts of sequence, duration, simultaneity, and age, and some of these experiments are described here.

Sequence and Duration

Earlier, in Part One, the means for assessing children's concepts of sequence and duration were described (refer to Figure 1–1). Two clear containers, one above the other, are used to depict a series of events that are then analyzed in terms of various interrelationships. The upper pear- or funnel-shaped container (that is, an Erlenmeyer flask) is filled with water, stoppered, and inverted with the necessary outlets and clamps to allow the water to drain into the lower container which is the shape of a slender cylinder with the same internal volume but taller than the upper flask. The child is supplied with drawings of the arrangement, and as the experimenter allows a certain quantity of water to flow into the lower container, the child is asked to draw the respective water levels at each of the six to eight intervals. Each drawing is cut so as to separate the two containers with their respective water levels drawn by the child, and questions are asked that are designed to reveal sequence and duration concepts. As reported in Part One, it was only after the age of eight or nine years that children in Piaget's study could relate all of the various events correctly in terms of sequence (order) and duration.

Before going to an analysis of what Piaget has referred to as physical time (as compared to psychological time), a statement by him on the importance of the concept of time in the total scheme of things is presented just as it appeared in *The Child's Conception of Time,* immediately preceding Part II, Physical Time. The words had a special meaning for me and perhaps will have for you, the reader.

> In short, much more even than qualitative time, quantitative time involves a system of geometry, a system of dynamics, and a system of mechanics all at once, since over and above the relation between the work and the rate at which it is done, it also introduces the constancy of the velocities under consideration (uniform rectilinear motion or regular periodicity). Time thus becomes an integral part of the structure of the universe. In fact, the four great categories of thought resulting from the use of infralogical operations, or operations in space-time, constitute an inseparable whole: object (or substance) and space; causality and time. For if it is true to say that there is no object without space, nor any space without objects, then it is also true to say that the interaction of objects defines causality, and that time is nothing other than the coordination of these interactions or motions. It is from causality that time derives its order of succession, because causes are necessarily anterior to effects, and it is causality which is expressed by duration, since duration is

simply the qualitative or quantitative ratio of distance to velocity (or, which amounts to the same thing, or work to "power").

These words will surely have different meaning to different individuals, but there is an unmistakeable interpretation of time as a central concept, interrelated with the whole of life.

Succession and Simultaneity

Physical time in our lives is usually concerned with discrete moving objects rather than the intermittent flow of liquids. Successive positions of the objects form a sequence just as the flow of liquid can form a sequence in terms of water levels. If two objects start simultaneously and stop simultaneously, we know that the durations are equal, but we can know nothing definitive about relative positions. Piaget used two mechanical toys, each of a different color, to determine how children conceive of succession and simultaneity. Toy A is capable of moving faster than toy B so that if toy B covered the distance x in a given time, toy A might cover x or $2x$ or $3x$ in the same time. By varying speeds, durations, starting points, and the like, and asking such questions as: Did they stop at the same time?, Which one moved longer?, Which one stopped first?, the experimenter can determine how the child views duration and succession, temporal and spatial succession, duration and velocity, succession and simultaneity, and so on. When the child is able to deduce one relation from another and realize that time is inversely proportional to speed (that is, a faster object requires less time to move a given distance) time and space become coordinated into a consistent whole.

In an interesting study related to the preceding one, Mori and Tadang (1973) projected motion pictures at slow, normal, and high speed with 42 eleven- and twelve-year-olds. The camera had been set at the center of a track with a 40-meter (125 feet) radius and followed a runner around the track. Each child was then shown the film on two different screens, one above the other. Each screen depicted exactly the same events, but at different speeds (that is, slow-normal, slow-fast, normal-fast). Fewer than half of the eleven- and twelve-year-old children were able to sort out the relevant data to correctly answer such questions as: Which runner went farther? and, Were the projected times the same?

These results would seem to suggest that a child's operational concept of time (duration and succession) is not achieved as early as reported by Piaget (about eight to nine years) but may be even later than the age of eleven or twelve years. In fact, Piaget (1946b) has reported that circular motion is more complex than linear motion and not until about the age of eleven years do children achieve a correct solution.

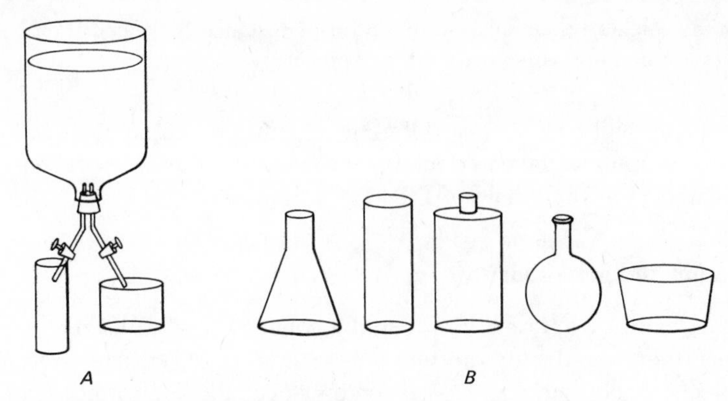

Figure 2–39. Y-tube apparatus for durations.

Another well-known experimental procedure used by Piaget and reported in *The Child's Conception of Time* involved a container that emptied liquid into two lower containers by means of a Y-tube (see Figure 2–39). Bottles of various shapes and sizes were used and children were asked questions about duration and volume and the like in an effort to determine a child's grasp of synchronous durations and the transitivity of time relations. It was found that by the age of eight or nine years, children were able to grasp the transitivity of equal or unequal durations or quantities. That is, if $A = B$ and $B = C$, then $A = C$ or, if $A > B$ and $B > C$, then $A > C$.

The outcome of the development of an operational concept of time is a readiness to quantify the relations in terms of measurements. Once the qualitative relationships are conceptualized, measurement or a quantification of the relationships can make sense to the child. The most tangible outcomes involved in science experiments usually require measurements of some sort. Unfortunately, children are often required to quantify the relations before they have operationalized the qualitative aspects, thus forcing a kind of rote learning process and communicating to the child that science is a rather mysterious process of measuring things that are unclear or even unknown.

Qualitative Speed

In *The Child's Conception of Movement and Speed* (1946b), Piaget describes the development of both qualitative and quantitative speed in terms of six operational systems:

1. *Placement* of objects, that is, order and succession.
2. *Displacement* or change of position.

3. *Co-displacement* or succession in time (duration) and speed.
4. *Relative displacements* or correlative movements and their speeds.
5. *Extensive* operations that permit ratios or proportions between time and distance.
6. *Metrical* operations that permit quantitative measurements of speeds.

The development of qualitative speed as a kind of "intuition" was assessed by Piaget using two tunnels of unequal length and two dolls with rigid wires attached to their backs. Both dolls enter and exit the tunnels at the same times and the child is then asked if one went faster. By about the age of seven years, most children indicated that they had an intuitive idea of speed by coordinating the duration and tunnel lengths. A similar test involved moving two objects along two paths with similar endpoints, but of different distances as in Figure 2–40.

The results from this experiment were similar to those from the tunnel experiment, with children of seven years of age or more saying that objects traveling along paths *A* or *C* would go faster than those moving along paths *B* or *D*. These findings are not in opposition to those described earlier with time concepts since co-seriation of durations, transitivity of durations, and the like are not required to determine relative speeds when time is the same and distance traveled is different.

A study by Raven (1967–68) tested 160 children in grades K–3 on a task that was almost identical to the tunnels and dolls task used by Piaget, except that cars were used and the tunnels were almost three times as long. Raven's results showed that only 43 per cent of the eight-year-olds were able to give correct answers with logical reasons to questions about the relative speeds of the cars. Raven's data differ considerably with what Piaget had reported, namely that most children by the age of seven years could relate the different tunnels with different speeds. It should be noted, however, that in a more recent replication study by Raven (1972) with the same number of children, 65 per cent attained a positive score by grade 3. This is a difference or variability of almost one half compared to the earlier results of 45 per cent and is in much closer agreement with Piaget's findings.

A final experiment on qualitative speed is described here before we turn to quantitative speed. It deals with relative speeds or the coordina-

Figure 2–40. Paths of different distances but similar end points.

tion of two speeds into a single speed. In reading Piaget's account of the experiment on relative speeds, one cannot help but see the influence of Einstein's earlier questions and his relativity theory developed in the very early 1900s. Eight cyclists were attached to an endless belt, one in front of the other, and a doll was on a separate belt parallel to the cyclists representing an observer (shades of special relativity!). At first the doll is stationary relative to the cyclists allowing all to pass it and then the doll is moved at constant speed, usually less than that of the cyclists. Each cyclist has a numeral attached to it so as to avoid confusion about how many cyclists will pass the observer in a given period of time. The various questions are intended to find out how the children can coordinate the two speeds when the two relative speeds are changed, even to the point of changing direction. From about the age of eight to eleven years, Piaget found that children learned gradually how to cope with the coordination of the two speeds, but only after some trial and error. Not until after the age of eleven years could the children predict correctly in advance of actually seeing the two motions what would happen when one speed was changed relative to another. It should be noted that only 50 children were tested or about five to six children per age level so the sample at representative ages was very small and no mention was made of how many children past the age of eleven years were unable to deduce relative speeds in advance of the actual experiment.

It appears that conceptualization of relative speeds, where two or more systems are involved, requires that one deal with each system separately in terms of distances, time, nature of the motion, perspective, and the like, while at the same time coordinating the various systems into an integrated whole.

Quantitative Speed

The mental operations that gradually develop and allow a child to understand speed (placement, displacement, co-displacement, and relative displacement) are followed by the development of quantitative operations where proportions are used in a measurement sense. Where two systems of movement (constant speed) are involved, distances and times can vary in the following ways:

1. equal distances, equal times.
2. equal distances, unequal times.
3. unequal distances, equal times.
4. unequal distances, unequal times.

The first of these, equal distances and equal times, is not a particularly interesting case but the remaining three cases offer opportunities to study the child's conception of speed. With cases 2 and 3, Piaget reported that children were unable to understand speeds of movements in succession until after the age of nine or ten years but that simultaneous movements of equal distances, unequal times, or vice-versa were successful one or two years before this. Raven (1972) supported this observation with 96 children in grades 3–6 using experimental interviews similar to those reported by Piaget (1946b). In both studies an object was started off at a right angle to a baseline, and the path (rectilinear) was sketched on paper and the time was recorded. Immediately after this another object traveled along a parallel line for the same distance, but the time was doubled. Each child was then asked if each object traveled at the same or at different speeds.

Case 4, where both distances and times are unequal, provides additional difficulties for children since the proportions involved are more complex. All parts of the proportions in distance and time now vary, and the child must be able to coordinate direct and inverse proportions simultaneously and successively, which amounts to a very complex task. It is not until after the age of ten or eleven years that children develop the necessary mental structures to allow for the solution of such complex problems.

A final example of the quantification of speed is described here as the conservation of uniform speed. The child is shown two parallel lines on a sheet of paper and an object moves along each line. The child is told that one object (*A*) moves a certain distance during a day and the other (*B*) moves only one half the distance in that day. Two questions are then asked: "How far will object *A* travel on the seventh day?" and "How many days will object *B* take to catch up?" Raven (1972) reported that only about one half of the 24 ten-year-olds he tested were able to respond correctly to the questions with logical reasons. His results correspond fairly well with those reported by Piaget. It is not until after the age of ten or eleven years that children can anticipate the regular increase in distance and determine the constant ratio between the two speeds.

It does not seem unreasonable to assume that measurement of speed or calculation of velocity can only make sense to a child who is able to conserve uniform speed. To expect such measurement before conservation has been achieved is equivalent to expecting a six-year-old to engage in symbolic arithmetic operations before conservation of number is achieved. Each child necessarily must engage in mimicry or rote memorization since understanding could not at that point be achieved. The relationships linking distance, time, and speed ($s = d/t$, $d = st$, $t = s/d$)

obviously imply proportional operations, and merely "plugging in" numerical values in no way ensures understanding. In fact, one of the greatest mistakes in science teaching that can be made is to communicate to students that "knowing" is equivalent to correct solutions of such "plug-in" exercises.

Acceleration

Since the conservation of speed implies an analysis of ratios between dissimilar uniform speeds, Piaget decided to extend the motion studies to acceleration. Uniform acceleration is really just a regular increase in speed (velocity) during a given time interval. Piaget (1946b), and later Raven (1972) with a replication study, analyzed the child's ability to conceptualize a regularly increasing velocity (acceleration) down an inclined plane that had flags placed at regular intervals along the edge. When the distances between the flags were equal and the children were asked to order the various speeds from slowest to fastest, by the age of nine to ten years most children were able to do so. However, when they were asked to order the times the car took in going from one flag to the next, very few nine-year-olds and only about one half of the ten-year-olds in Raven's study were successful. Even fewer children could order the distances when time intervals were equal (40 per cent by 10 years).

Whereas an intuitive concept of acceleration, that is, a realization of constantly increasing speed, is apparently achieved by the age of nine to ten years, it is not until two or three years afterward that children can coordinate the times, distances, and speeds in a consistent manner. Many physics teachers, including this author, at the secondary and postsecondary levels will readily agree that acceleration is not well understood even by many in the sixteen- to twenty-year-old group. Going from a qualitative to a quantitative or a metrical analysis of acceleration apparently causes very difficult problems for many individuals past the age of eleven or twelve years.

Mass, Weight, and Density

Like time and motion, concepts of mass, weight, and density permeate the science textbooks and curriculum projects in use today. In fact, matter-energy equilibrium models are the central focus in many science curricula, both at the elementary and secondary levels. Even at the most elementary levels, there is the implicit assumption of conservation of matter, and often, the further assumption that children can conceptualize "energy" in much the same way they grasp concepts such as soft, hard, rough, and wet. We look at a number of experiments by Piaget and others in this section on the child's development of matter-related

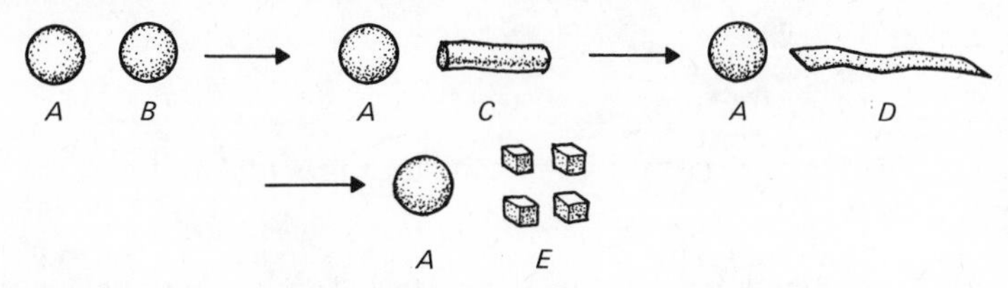

Figure 2–41. Clay balls used in mass, weight, and volume conservation.

concepts and how they might be related to other mental structures we have already considered.

Piaget and Inhelder (1941) conducted many experiments using clay as the substance for determining conservation abilities of children up to about twelve years of age. Each child was shown two identical balls of clay and the shape of one of the balls was then changed into other shapes as shown in Figure 2–41.

Questions such as "Which ball has more clay?" and "Do they take up as much room?" were asked of each child. Piaget and Inhelder found that, in general, conservation of matter occurred at about the age of nine years, conservation of weight at about the age of ten years and conservation of volume at about the age of twelve years.[3] In a replication study, Elkind (1961) confirmed these results for mass and weight using similar materials and techniques but in a more recent study with 250 children, Good et al. (1973) found that weight conservation is achieved by only about 30 per cent of children in fifth grade (ten to eleven years of age). Aluminum foil rather than clay was used in this study and that, apparently, was the major difference. As with previous differences noted in area and volume experiments, this rather substantial difference on weight conservation serves to emphasize the importance of the materials used during any interview situation. Although it may not be readily apparent to the adult, even subtle changes in the interview situation are often enough to elicit significant changes in children's responses.

A recent study by Good (1973) yielded rather surprising results on a paper and pencil conservation-of-weight task. Figure 2–42 shows the drawings and possible choices on the weight task with the college seniors.

1. *A* is heavier than *B*.
2. *A* is more dense than *B*.
3. *A* is attracted more strongly by gravity than is *B*.

[3] These are compromise figures taken from two well-known authorities in the field, David Elkind and John Flavell. Elkind (1961) has used the ages of 7–8 years, 9–10 years, and 11–12 years for mass, weight, and volume, respectively. Flavell (1963) has used the ages of 8–10 years, 10–11 years, and 12 years and after in interpreting the results reported by Piaget and Inhelder (1941).

4. Either 1 or 2.
5. None of the above.

The following question corresponded to the drawings and choices seen in Figure 2–42:

Two pieces of aluminum foil are 4 inches on each side and are of equal thickness. *A* is folded in half four times so that each side is 1 inch in length and it is 16 times as thick as *B*. When the edges of *B* are turned up slightly, it floats on water. When *A* is gently dropped into the water it sinks to the bottom of the vessel. To explain this, one might say:

The results of this item point up the difficulties inherent in designing valid paper and pencil measures for developmental abilities. For those readers who are conservers of weight, it is easily seen in Figure 2–42 that "none of the above" is the best choice since none of the previous four choices explains the behavior of the two pieces of foil. However, only 9 per cent of the 98 college seniors chose "none of the above," instead preferring the second choice, "*A* is more dense than *B*" (69 per cent). Other work by the author on this task with college seniors and high school students caused him to anticipate that about one third of the students would indicate a nonconserving response. Since this figure was actually more than two thirds, it was obvious that something else was being tested. Apparently, it was the students' understanding of density, or more accurately, a lack of such understanding, that was really responsible for such a high percentage of nonconserving responses. Most students believed that folding the aluminum foil actually changed its density.

Piaget and Inhelder (1941) reported results of experiments that were specifically designed to assess a child's concept of density, using popcorn. Until after the age of twelve years, children thought that when the kernel popped, it was heavier. Since, by definition, density assumes an independent relationship between weight and volume, it would not seem possible that a child could conceptualize density before understanding the concepts of weight and volume. This assumption seems to be borne out by the data and, in fact, suggests that density is not conceptualized very well even by adults. The development of an operational concept of density apparently assumes a genuine understanding of the

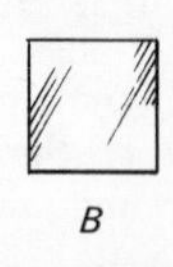

1. *A* is heavier than *B*.
2. *A* is more dense than *B*.
3. *A* is attracted more strongly by gravity than is *B*.
4. Either 1 or 2.
5. None of the above.

Figure 2–42. Paper and pencil "interview" for conservation of weight.

particulate nature of matter so that small, heavy objects or large, light objects can be compared and incorporated into an overall system of matter states. The atomic-molecular model is just that, a model. It assumes that a person can conceptualize this rather abstract idea and, in matter-energy relationships, equate "nonmatter" to matter that can be seen and touched.

An interesting study begun by Flood (1974) attempted to identify matter change concepts held by high school and college students. Flood attempted to look into student's conceptions of matter changes that are not visible as is usually the case in most physical, chemical, and biological changes. The main system used in that study was a chicken egg, and questions were directed to possible changes in matter during embryo development and hard boiling. Such changes are obviously more abstract than clay ball transformations, which we have seen are not conceptualized successfully until after the age of eleven or twelve years. Depicting energy as a "special form" of matter and thereby expecting matter conservation with physical, chemical, and biological systems for elementary and even many secondary students is likely to result in an empty kind of rote learning. Very few studies have been directed toward a greater understanding of how children develop a conception of the rather abstract idea of matter conservation beyond the visible deformation experiments with clay, water, and the like. Those who have tried to "teach" some form of matter-energy conservation are well aware, at least in a subjective sense, of the lack of genuine understanding by students. Well-designed studies are needed to provide "objective" evidence for the feasibility of teaching such abstract concepts to students at post-elementary school levels.

A final study on weight is briefly described here from a somewhat different viewpoint. Ninety college seniors were asked to decide how an equal arm balance (see Figure 2–43) could be "balanced" using available weights (masses for the physicists).

Students were asked where the 20 unit weight should be placed to balance the 5 and 10 unit weights. Only 20 per cent of the 90 students replied correctly that it should be placed on hook 3. It can be seen later

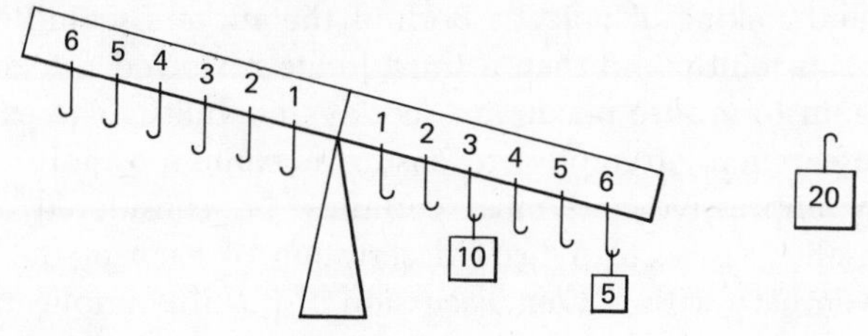

Figure 2–43. Equal-arm balance.

that although weight is involved, the fact that so few of these adults could respond correctly cannot be attributed solely to a faulty conception of weight. Equilibrium in the balance involves the simultaneous coordination of two variables, which is more a test of formal operational or abstract thinking than concrete operational thinking. Tests for formal operational thinking are considered in some detail in a later section since post-elementary school science very often requires such advanced thought.

Causality

Before going on to an analysis of formal operational thinking, it seems appropriate to mention a few characteristics of the development of what could be called "causal" thinking in children. Science seeks to explain various natural phenomena in terms of what makes sense to the "observers" at a particular time. The observers are usually called scientists, and more often than not what makes sense to them may not make sense to others not engaged in similar work. Thus, explanations (theories) for causality that make sense to children will usually bear little resemblance to the abstract, formalized explanations of causality offered by "scientists." In a work that is now about 50 years old, Piaget (1927) described the ideas of children from about the ages of four to twelve years on their conceptions of physical causality. There have been many criticisms of that particular set of studies and the corresponding ideas about "stages" of development, and a more recent study (Laurendeau and Pinard, 1962) attempted to overcome some of these criticisms by specifying sample size, characteristics, and other specifics. The latter study was truly comprehensive and was designed in such a way so as to avoid the criticisms that had been associated with Piaget's earlier work.

The sample was composed of 250 girls and 250 boys between the ages of four and twelve years. The complete project used 27 distinct questioning areas but only 5 are reported in the present work. The total time devoted to the examination of *each* child was about 10 hours, distributed over 4 to 8 sessions, which amounts to a massive total of 5,000 hours! The five areas of causal thinking examined were: (1) concept of dream, (2) concept of life, (3) origin of night, (4) movement of clouds, and (5) floating and sinking of objects. Both of the authors scored each of the 2,500 protocols jointly and then a third judge reviewed the entire 2,500 protocols again to ensure maximum scoring reliability. The extent to which these researchers attempted to ensure maximum validity and reliability is truly impressive and must certainly be considered as a "standard" in causality research. A brief description of each of the five causal areas is given here with a later discussion of possible implications beyond the elementary school level.

Concept of Dream

Such questions as: Do you know what a dream is?, Where does a dream come from?, and Who makes the dreams come? were asked of all the children and their responses were recorded verbatim.

The authors found four distinct stages or categories of ideas about the nature of dreams:

1. Incomprehension or Refusal

   About 50 per cent of the four-year-olds and 25 per cent of the five-year-olds made up this group. They either refused to answer or gave no indication of understanding the questions.

2. Integral Realism

   This stage is characterized by statements that the dream is external to the dreamer (on the wall, window, pillow, and so on) and the cause is God or the devil or the sandman, or other supernatural figures. About 25 per cent of the four- and five-year-olds and 10 per cent of the six-year-olds fell into this category.

3. Mitigated Realism

   During this stage, the children progress toward viewing the dream as being entirely interiorized and seen only by the dreamer. However, they begin by "seeing" the dream as originating on the inside (head, heart, eyes) but taking place on the outside. Some children view the process in reverse order with the dream originating on the outside and going inside for viewing purposes. Another characteristic of this stage is that the children believe that the dreamer is not the only one who can look at the spectacle. This suggests an interpretation of dreams as materially real events rather than as non-materialistic ideas generated by the dreamer. Even by the age of seven to nine years the authors found that about 20 per cent of the children were in this stage and at the age of five and six years it was about 40 per cent of the children.

4. Integral Subjectivism

   During this stage, the remaining traces of realism disappear so that the origin of dreams is solely within the person and no one else can "see" the events. After ten years of age, only about 5 per cent of the children of the 500 interviewed gave responses that indicated they had not yet achieved this "final" stage.

Concept of Life

In Part One, animism was referred to as attributing life to inanimate objects. By asking questions about the nature of "aliveness" and the living qualities of various objects (for example, sun, table, cat, cloud, lamp, fly), Laurendeau and Pinard were able to categorize different types of animism and, finally, identify those children who were beyond animistic explanations. The results are rather surprising for they show that even by the age of eleven or twelve years many children still use animistic thinking as a part of their explanations about life concepts. Up to about the age of six years, a majority of the children do not understand the meaning of the questions and respond by giving reasons that are constantly contradictory (that is, using the same explanations for both life and nonlife) or by answering "yes" or "no" indiscriminately. From six to nine years of age, children use a type of animistic thinking, or more often, a combination of types such as movement (clouds) or usefulness (sun) to categorize objects into living-nonliving groupings. Table 2–16 shows total percentages of errors for each inanimate object discussed with all children. Interestingly, the four "heavenly" objects top the list of inanimate objects that are "given life" by the children. Also, even though an age distribution is not presented in Table 2–16, the authors did include such data in the original work and beyond the age of five or six years, there is not much variation in the percentage of animistic errors among age levels for the "heavenly" objects. In fact, for the remaining 10 objects, there was little variation in the percentage of errors for children in the age group nine to twelve years. The twelve-year-olds made about as many errors as did the nine-year-olds and in some cases, more (for example, fire, 25 per cent to 15 per cent).

TABLE 2–16. **Percentages of Animistic Errors for the 14 Inanimate Objects Considered by the 500 Children (Laurendeau and Pinard, 1962)**

Object	Order of Presentation	Overall Per Cent in Error
sun	2	43
wind	9	35
cloud	5	29
rain	12	25
car	4	25
fire	11	23
plane	10	23
bike	13	20
watch	7	19
bell	8	19
mountain	1	14
lamp	6	13
pencil	14	12
table	3	5

The results of this study of children's conceptions of life most certainly have implications for curriculum and instruction decisions at the elementary school level. Although we found in earlier sections on classification and seriation that children often developed concrete operational abilities by a given age somewhere in elementary school, it seems that some concepts (for example, life) are dealt with in an inconsistent manner even by a large portion of eleven and twelve-year-olds.

Origin of Night

Artificialism, or the concept that there is a "maker" at the origin of all things, was the most common kind of explanation by children to questions about the origin and nature of night. Such questions as What is night?, Why is it dark at night?, and Where does the dark come from? were used with the children and until six years of age, about 75 per cent gave answers that were considered to be either "absolute artificalism" or incomprehension. Dark is made "so we can sleep" or "so we know when to go to bed" and God is the originator of the dark and of night.

A middle stage combines artificialistic thinking with physical interpretations of dark where the sun or clouds or smoke are somehow involved, but God is still involved in many cases. By the age of ten years more than two thirds of the children explain the darkness of night in terms of physical and natural events although traces of artificialism still exist in some children.

Movement of Clouds

Until the age of eight or nine years most children attribute cloud movement to God or a Blessed Virgin or the moon, or sometimes children say that the clouds just "move by themselves." By the age of nine or ten years, a majority of the children have decided that the wind moves the clouds but quite often a bit of precausal thinking creeps into their explanations.

Floating and Sinking

This interview is different from the previous four examples in that manipulative materials are used as an integral part of the procedures. Cylinders, plasticene, marbles, and similar objects were used and the children were asked if they thought each object would sink or float and why. Virtually none of the children, even by the age of twelve years, were able to formulate the correct explanation for sinking and floating, that is, comparative densities *and* shape, although a majority by the age of eleven and twelve years used physical rather than precausal explanations (for example, artificialism, animism).

This particular interview procedure was used by Piaget during the mid-1920s and again later for a purpose different than determining causal explanations. We see in following sections how Piaget used this and other interviews to study the thinking mechanisms involved with logical thought.

Although the Laurendeau and Pinard study does not attend to the growth of logical structures, per se, it does have great relevance for science teaching. Causality or explanations for physical phenomena are central to the work of scientists, and the resulting theories and models that turn up in various forms in public school curricula assume causal rather than precausal thinking. Darwinian evolutionary theory can make little sense to the student who uses artificialism in explanations about the origin of things.

Summary of Development at a Concrete Level

Development toward logical thinking is a slow process involving the active participation of the child. Piaget and his colleagues in Geneva are obviously the leaders in the field of experimental research into children's thinking. There are, however, sufficient variations between their findings and those of many other workers in the field that a summary must reflect the variations. Therefore, in the age ranges associated with the numerous concepts and processes that follow, some ranges are in agreement with those reported by Piaget and others in the Genevan group, whereas others vary somewhat. The variance is usually such that ages associated with concepts are greater so that a wider range is indicated. Area conservation, for example, has been reported by the Geneva group to be attained by the age of seven to nine years. Many other studies differ with the range by a couple of years and consequently the summary that follows reports a range from the age of eight to eleven years.

It is very important to emphasize that the age ranges are *approximations* that will include most but not necessarily all children within that range. Cognitive development is a *continuous* process that begins at birth, or before, and shows a remarkably similar sequence among individuals. There seems to be a logical or preferred sequence, so to speak, in the development toward logical or "formal operational" thinking. Some children might develop a particular cognitive ability before the "typical" age range or after that range, but the *order* of this development seems to be invariant, at least in a general sense. Most of the age ranges that follow fall within the overall range between the age of six or seven to eleven or twelve years. This level of development corresponds to Piaget's "concrete operational" stage of development, but it is referred to here merely as concrete development. In experimental and theoretical science it is the later abstract or "formal operational" thinking that usually is required.

Concrete Development

	Attained at Age Range
Classification	
a. Intensive–extensive coordination	7–9
b. Stable classification criteria	7–9
c. All–some relations	8–10
d. Class inclusion	9–11
Seriation	
a. Simple series	5–7
b. Insert elements into series	7–9
c. Analyze elements using whole series shape	9–10
Number	
a. Conserver	6–8
b. Coordination of ordinal and cardinal properties	7–9
c. Singular class	8–9
d. Null class	9–11
e. Multiplicative classification	10–12
Topology	
a. Rhombus reconstruction	6–7
b. Linear and circular order	7–8
c. Knot tying	8–10
d. Continuity	10–12
Projective Geometry	
a. Straight-line construction	7–8
b. Perspective	
Simple objects	8–9
Complex structures	10–11
c. 3-D Visualization	
Simple structures	7–8
Complex structures	10–12
d. Shapes from cutting solids	
Simple	7–8
Complex	11–12

	Attained at Age Range
Euclidean Geometry	
a. Water-level conservation	9–11
b. Conservation of parallelism	9–11
c. Conservation of angles	10–12
d. Metric proportions	12–16
Length and Distance	
a. Conservation of distance (empty space)	7–8
b. Conservation of length (filled space)	6–8
c. Iteration of length	8–9
d. Subdivision of straight lines	
Staggered but equal	7–9
Staggered but unequal	8–10
Area	
a. Conservation of area	8–11
b. Measurement by iteration	8–11
c. Measurement by superposition	9–11
d. Perimeter–area relationship	12–
Volume	
a. Conservation of internal volume	9–11
b. Conservation of displacement volume	11–15
Coordinate Systems	
a. Generalized reference system	11–14
b. Angle measurement	10–12
Time and Motion	
a. Co-seriation of sequence and duration	8–10
b. Succession and simultaneity	8–10
c. Transivity of equal or unequal durations	8–10
d. Time measurement	
Linear motion	9–10
Circular motion	11–14
e. Relative speed	11–14
f. Conservation of uniform speed	10–12
g. Acceleration	12–
Mass, Weight, and Density	
a. Conservation of continuous quantity	8–10
b. Conservation of weight	10–12
c. Density	12–
d. Weight–distance equilibration	12–

The attainment of an operational understanding of the various concepts previously identified is usually realized within the indicated age ranges. In reality, of course, there is no abrupt beginning or ending. The process is gradual and even for concepts that appear to be "clear-cut" and easily definable such as conservation of number, area, and weight, wide variations can be traced to factors within the experimental interview procedures. The nature of the materials used, type and number of questions asked, sequence, criteria for acceptable responses, and other aspects all can have an important influence on the "results" of the interviews. The level of difficulty of concepts also can be incredibly different. Motion, for example, can be the simplest movements perceived by the very young infant or it can involve very complex relativity theory describing and predicting hypothetical relative motions. From the ages of six and seven to eleven and twelve years, the child is developing a "groundwork" of sorts that is prerequisite to, and necessary for, abstract thinking of the hypothetical-deductive sort. Conservation continues to develop as a useful concept and is not something "attained" at a particular age level. Conservation of number assumes the ability to "conserve" or to hold invariant the number of elements in a given set regardless of any spatial reorientations that occur. This ability usually develops between the ages of six and eight years but in no way assures conservation of area or weight, or the like. As a matter of fact, the conservation of matter (eight to ten years) or of weight (ten to twelve years) in no way ensures the attainment of conservation of matter at molecular, atomic, or nuclear levels. Conservation of energy or matter-energy equivalency appears regularly at the early elementary levels in textbooks and other curriculum materials in spite of the fact that matter or weight conservation is not attained until mid- or late-elementary school levels. And this type of conservation is at a concrete level where the child can see and touch the material being transformed, not at the abstract, hypothetical level of conservation of energy or matter in chemical or nuclear reactions.

Summarizing a "concrete" level of cognitive development can easily be interpreted as an end to a "stage" and, of course, that is the very thing that I have attempted to avoid in this summary. A "beginning" or an "end" to a stage is really only an arbitrary definition by persons attempting to communicate with one another, but does not reflect the true nature of the developmental process.

Reversibility is also identified by Piaget as a very important characteristic of concrete operational thinking. The ability to mentally reverse a process, such as moving objects around in a conservation of number interview, allows the subject to see that nothing has really changed. One might conclude that the child now has reversible thinking at her/his disposal, but this is not really the case. If it were, then the subject would be able to apply that same thinking to other conservation tasks, but we have already seen that such is not the case. Reversibility as a useful cognitive process continues to develop to the extent that it can gradually be applied to other, more complex concepts. No abrupt beginning or ending is associated with the ability, just a continuous development.

The next few sections on probability, combinations, proportions and the like fall under the heading, "Concept Development at an Abstract Level." Just as there is no abrupt ending of concrete development, there is no sudden beginning of abstract development. It is merely a continuation of the *same* developmental process, influenced by a multitude of factors, such as overall physical health, opportunities for appropriate cognitive experiences, emotional equilibrium, and feelings of being needed and loved. When these and other factors influencing the cognitive development of a human being are operating at an optimal level, the rate of development of mental structures required for logical thinking is also at a high level.

concept development at an abstract level

Introduction

The material in this section has been included for the reader who is interested in abstract or formal thought, as defined by Jean Piaget. It does not have as much relevance to the elementary schoolchild as the previous material and so can be considered *optional* if the reader's interest in cognitive development is limited to grades K–6. Particularly for the sections titled *Propositional Logic* and *The INRC Group*, the material becomes rather involved.

The Summary compares formal and preformal thought, and the reader who is not very interested in more theory and details may skip immediately to that part. Even for the elementary schoolteacher, however, a fairly good understanding of the characteristics of abstract or formal thought is valuable. A large proportion of many elementary science textbooks often presents science concepts that require a type of abstract thinking which is simply not available to the students. This problem is dealt with in Part Four in some detail as a practical application of having a functional understanding of the characteristics of abstract and concrete thought.

The development of an operational understanding of concepts of classification, seriation, number, space, time, weight, and others that we have considered thus far are necessary prerequisites to later abstract thinking. What has been described under the major heading of "Concept Development at a Concrete Level" corresponds to what Piaget has called "concrete operational" thinking. It is misleading, however, to think of a generalized type of stage beginning all at once. Conservation or the ability to maintain permanence of quantities and events in spite of perceptual (spatial) changes is developed by children according to the concept in question. In general, number conservation develops before area conservation, area before weight, weight before volume, and so on. Even this type of sequence can be an overgeneralization at times, however since we have seen that there are different levels of difficulty of area conservation or of weight conservation. There does seem to be a common characteristic among the many types of thinking used by children from the age of six or seven years to eleven to twelve years of age. Children during this stage of mental development must begin with real objects and events and are unable to deal with hypothetical possibilities. Problem solving proceeds from present data, and possible solutions are limited to the physical properties of the immediate objects and events. This general characteristic is, of course, a major limiting factor for children in elementary school. Any concepts or theories that cannot be approached by the children in a concrete way, where their own actions on objects provide the primary pathway for understanding, must be considered as inappropriate for the curriculum. Let us now look at some concepts and corresponding cognitive strategies that characterize formal operational thinking.

Chance or Probability

The concept of chance or probability is important in those areas in science where large numbers of objects or events occur and some type of quantitative interpretation is desired. In some cases, new branches of study have evolved that are based on the concept of chance or random events. Quantum mechanics in physics, genetics in biology, and models for weather prediction in meteorology are all based on probability models. In a way, much of the problem solving that occurs in our day-to-day lives is linked to making determinations about the chances of something or other happening. Studying the history of related events and

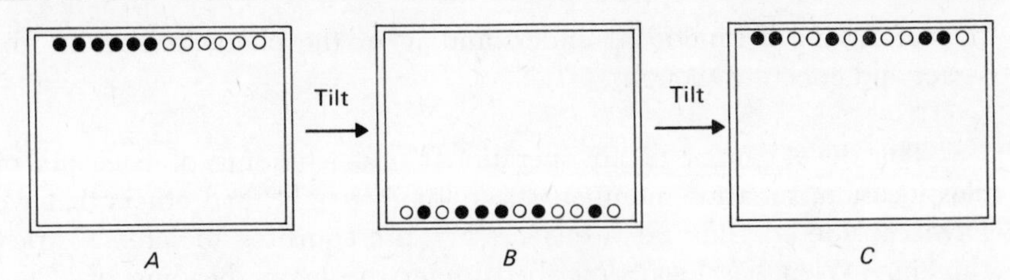

Figure 2–44. Experiment with red and white beads used for chance events.

applying "trends" to current problems is a kind of probabilistic thinking. This type of thinking was studied by Piaget and Inhelder (1951) and summarized later by Flavell (1963) in his comprehensive analysis of Piaget's work. A basic premise was used by Piaget; namely, that before people can conceptualize chance events they must first have developed the means to organize and predict the orderly, nonchance portions of the environment. One experiment used red and white beads arranged along one end of a rectangular tray so that all the red beads were on one half of the tray and the white beads on the other half. The tray was then tilted, which caused the beads to run to the other end, with some mixing in the process. (see Figure 2–44).

Further tilting, of course, causes greater randomization and the child is asked to predict each outcome by drawing the arrangements. As children approach the age of ten or eleven years, they become progressively more detailed and accurate in predicting and representing the randomness.

In another experiment by Piaget, children were asked to predict the shape of a distribution resulting from random movement of objects (Figure 2–45).

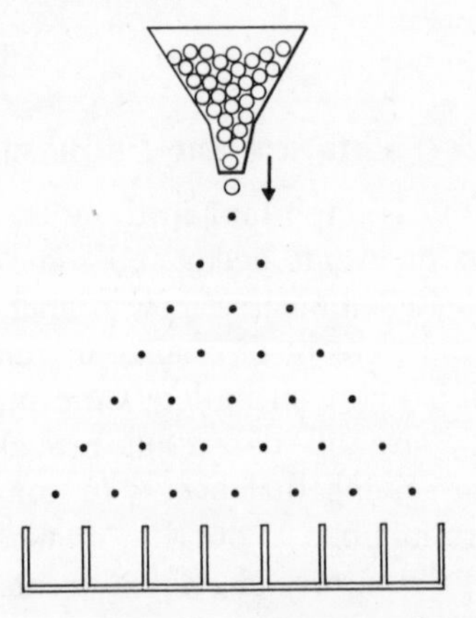

Figure 2–45. Apparatus for a random distribution.

Only after the age of ten or eleven years were the children able to predict with any accuracy that the distribution would become progressively more symmetrical as the number of marbles increased. Other experiments involved the drawing of various counters from a bag that contained unequal numbers of the counters. Children were asked to predict the likelihood of drawing a particular counter (color). With these and many other experiments, Piaget concluded that it is not until after the age of ten or eleven years that children are able to apply any kind of systematic strategy in attempts to predict outcomes with random events.

In an interesting study investigating the so-called "gambler's fallacy," Cohen and Hansel (1955) asked twelve- to fifteen-year-olds to predict outcomes in a coin-tossing series, given that previous x tosses had yielded a run of either heads or tails. Fifteen per cent of the twelve- and thirteen-year-olds and 27 per cent of the fourteen- and fifteen-year-olds were not misled by the "run" idea, instead expressing the belief that the next outcome was independent of previous outcomes. All of the other children operated under the assumption that past events (tosses) would in some way affect the next toss. Any gambling house would love to have adults with such views frequenting their establishments (and undoubtedly do).

A more recent study with high school biology students (Emery, 1973) used a 32-item test to determine probability reasoning. Each item described a "raffle" drum with contents such as black and white marbles and students were told that a scoop of the drum's contents yielded a certain number of marbles of each color. Students also knew the original contents of the drum, namely, 6,000 black marbles and 4,000 white marbles. Each of the 90 students was asked to predict what color marble he or she thought would be pulled out if the student then reached into the drum for one more marble, and why. Fewer than half of the 90 students consistently predicted that the next marble (object) should be whatever the majority color was in the drum. Over two thirds of the remaining students predicted that the color would make the ratio of scoop contents more like the ratio in the drum. This strategy of ratio matching as well as other misconceptions regarding the occurrence of chance events have been reported by Davies (1965), Gratch (1959), Jarvik (1951), Ross (1958), and many others. Young children very often believe that events such as drawing two colors of marbles from a sack should take turns, especially when the ratio of the colors is not too far from 1:1; for example 1:1.5 or 1:2 rather than 1:10 or 1:20. Though this belief may seem amusing to adults, the gambler's fallacy mentioned before can lead to predictions that are just as wrong. Chance events such as the roll of dice do not "remember" what has occurred before just as they do not politely take turns in some way. To the large majority of children and a smaller, but well-established, majority of adolescents and probably adults, concepts of chance or probability offer difficulty usually exemplified by ratio matching or some form of gamblers fallacy, also known as negative or positive recency. To expect such individuals to understand concepts in genetics or any other content based on probability theory is obviously naïve and possibly detrimental. The approach must be concrete and the time required for real understanding would undoubtedly vary considerably depending on the individual. A further complicating factor is that science concepts such as inheritance characteristics and quantum theory invariably involve more than one mental operation. Using genetics as an example, we see that other mental operations, such as using proportions and making combinations, are involved, in addition to probability concepts.

Inhelder and Piaget (1955) identified eight "notions" or formal operational "schemata" as Piaget has referred to them. Probability was among these eight notions, and both authors noted that successful problem solving with chance concepts is dependent upon a complete combinatorial system. At the concrete level, children are limited to additive composition to determine possibilities and it is shown in the section on combinations that some possibilities are omitted in the process. The authors also noted that the isolation of a law of chance required an understanding of correlations. A problem was set up using 40 cards, each with a face drawn on it. The variables were eye color (blue or brown) and hair color (blond or brown) as seen in Figure 2–46.

Each child was given a set number of cards and asked if he or she thought that there was a correlation between eye and hair color. Simple probability could be determined by discovering the ratio between a specific case, say 6, and the total number of cases, $a + b + c + d$. If there were 5 cards with blue eyes and brown hair, the ratio would be 5 to 40 or 1 to 8. The probability that you would get a card b from the deck of 40 is put at 1 chance in 8. Correlation, however, is a function of the difference between two cases divided by the total number of cases, that is, $a - b/a + b + c + d$. As the difference between the two cases decreases, the correlation coefficient also decreases. Inhelder and Piaget found that whereas many children by the age of twelve or thirteen years could determine the chances for blue eyes and blond hair by comparing them to the total number of cases $(a + b + c + d)$, most were still unable to compare (correlate) the chances. Not until the age of fourteen to fifteen years did the subjects interviewed give indications that they could find relationships between the ratios of simple chance events.

The eight "notions" or formal operations referred to earlier include, in addition to probability, combinations, proportions, coordination of

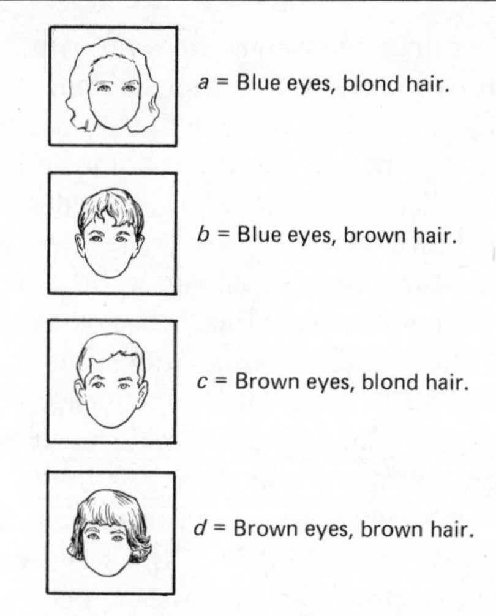

Figure 2–46. Cards used for correlation experiments.

reference systems, mechanical equilibrium, correlation, multiplicatve compensations, and nonempirical conservation. A brief look at some of these will lead us to propositional logic (16 binary operations) and how, what Piaget has called the INRC group, transforms them into a formal operational system.

Combinations

The process of dealing empirically with combinations of factors in science investigations is basic in real experimental situations. Identifying and controlling experimental variables in a systematic way assumes an underlying combinatorial system. Inhelder and Piaget (1955) used a clever experiment to determine the ability of children and adolescents to use a complete combinatorial system (see Figure 1–6 in Part One). Four containers with colorless liquids (sulfuric acid, water, hydrogen peroxide, and thiosulfate) and a smaller container of potassium iodide were made available to each child after a brief demonstration. Dilute sulfuric acid + hydrogen peroxide + potassium iodide will produce a yellow color, and the child is shown that a color change occurs through mixing but it is left for the child to determine what combination(s) will cause the change. A complicating factor is that the thiosulfate will prevent the color change and eventually this must be discovered. Up to the age of seven or eight years the children's actions were mostly random with only accidental successes. From the age of seven to nine or ten years many of the children began to associate one liquid with one other liquid but combinations that were more complex than one-to-one were not tried. It does not occur to the child to mix three liquids except in a random fashion, that is in the process of mixing all of them together to "see what happens." When the interviewer suggests that mixing three liquids might produce the desired color, the child does not follow up the idea by trying all possible three-way combinations.

By the age of ten or eleven years, Inhelder and Piaget reported that the children began to use more complex combinations, but not in any systematic way. They tried some of the possible combinations but their lack of a systematic approach caused some possibilities to be left out. The formation of a systematic combinatorial approach appears at the age of twelve or thirteen years. The usual pattern of mixing is to try all of the possible combinations of two, then with three, and so on until all groupings have been exhausted. Another characteristic of this new stage is that the child, or perhaps more accurately, the adolescent, does not stop with the investigation when a color change occurs. All combinations must be tried before the subject is satisfied. This is a significant sign that the subject now has the mental structures to go beyond the here and now to the "what is theoretically possible" domain.

Inhelder and Piaget identified a "final" stage that they called "equilibration" of the system in which the subjects reflected a systematic and thus faster approach to a complete solution with all possible combinations. When asked for a "proof" of a particular statement by the interviewer, subjects in this advanced, formal stage could quite easily offer direct evidence by constructing an appropriate combination.

Other more recent studies on combinatorial abilities of post-elementary school students have suggested that many, perhaps a majority of, such students are unable to accurately generate combinations and permutations. Smock and Belovicz (1968) reported that even high-ability junior high school students formed less than half of the potential combinations that could be made in certain tasks designed to assess such ability. They concluded that the "typical" junior high school student did not have an understanding of the principles of making combinations.

In a more recent study, Emery (1973) developed a test on forming combinations and administered it to 90 high school biology students. Three questions were involved:

1. Write all of the combinations of one letter and one number that can be made from these letters and numbers.

   A B C D For example, 1*A* is the same combination as *A*1.

   1 2 3 4

2. Write all of the combinations of one shape that has a letter inside with one shape that has a number inside.

[A] [B] [1] [2] For example, two combinations are

(A) (B) (1) (2) [A] [1] or (A) [1]

3. Next write all the combinations of *two* letters and *two* numbers which can be made from the letters and numbers below, according to this rule: in each combination, one of the letters must be inside a circle and the other one must be inside a square; one of the numbers must be inside a circle, and the other one must be inside a square.

[A] [B] [1] [2] For example,

(A) (B) (1) (2) [A] (A) [1] (2)

The results of the testing are shown in Table 2–17. It can be seen that only one student in five was able to pass all three items and only two in five passed items 1 and 2. Science content that assumes an ability to form combinations, such as in genetics, would certainly prove exceedingly difficult for high school students of "average" ability. Emery's study also assessed the extent to which direct instruction in genetics affected the students' abilities to form combinations. She found that direct genetics instruction had no significant effect on students' abilities to form combinations. More is said later of the attempts to "train" children to develop cognitive abilities, but for the moment it should be emphasized that what might seem to be a natural step to take in helping a child become more "knowledgeable," that is, direct training, could yield insignificant cognitive gains and would constitute an inappropriate instructional strategy for certain long-term goals, for example, independent thinking, creative problem solving, open-ended investigation, and seeing "conclusions" as tentative.

Before leaving the concept of combinations to look at other learn-

TABLE 2–17. **Results of the Test on Forming Combinations for High School Biology Students**

Achievement Level	Frequency	Per Cent
Pass 0	11	12
Pass Item 1 only	24	27
Pass Item 2 only	1	1
Pass Items 1 and 2	36	40
Pass Items 1, 2, and 3	18	20
	90	100

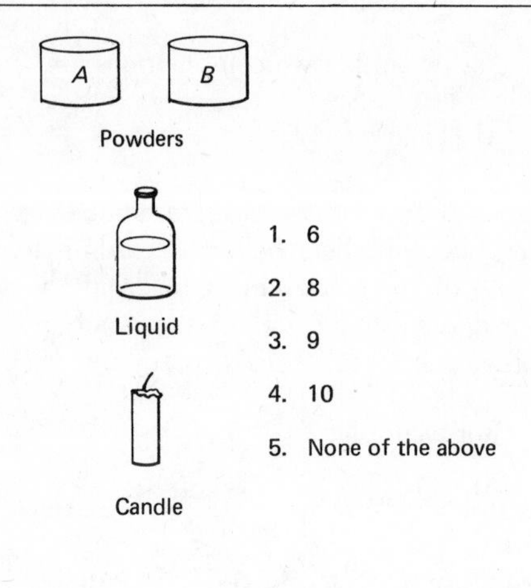

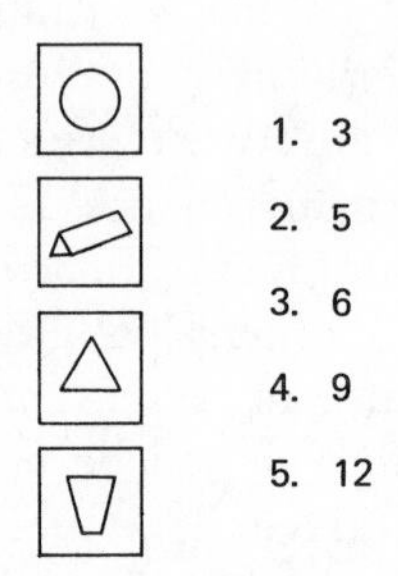

Figure 2–47. Two questions on making combinations.

ing areas requiring logical or formal operational thinking, the results of a study involving college seniors should be considered (Good, 1973). Two questions were presented to the subjects in written form. Drawings representing the objects referred to in each question accompanied the questions. Figure 2–47 shows the two questions as presented to the subjects. A resource person was available in each of the six classes in which the questions were administered to answer individual questions or otherwise assist the students.

Table 2–18 shows the results on each of the two questions.

TABLE **2–18. Results on Combinations Questions with College Seniors**

Choice	1	2	3	4	5
Question					
I.	18 (19%)	8 (8%)	31 (32%)	6 (6%)	33 (34%)
II.	0	3 (3%)	71 (73%)	4 (4%)	20 (20%)

Question I was considerably more difficult for these college seniors with only 32 per cent correctly choosing item 3, 9 combinations. Question II requires only a partial combinatorial system whereas question I requires a more fully operational system. The nine possible "tests" that can be performed on the powders are three:

1. $A +$ liquid.
2. $A +$ candle (heat).
3. $A +$ liquid $+$ candle.
4. $B +$ liquid.
5. $B +$ candle.
6. $B +$ liquid $+$ candle.
7. $A + B +$ liquid.
8. $A + B +$ candle.
9. $A + B +$ candle $+$ liquid.

An analysis of the responses to the question Why? eliminated the possibility that some students were misinterpreting the question. It would be reasonable, for instance, to assume that the *order* of testing might make some difference, for example, $A + B +$ liquid $\neq A +$ liquid $+ B$. This type of thinking was not confirmed, however, for students who chose the "none of the above" response.

By chance alone, it could be assumed that 20 per cent of those responding to such a question would choose the correct item. Taking this into account, the 32 per cent figure for correct choices might be further reduced. Although the depth of investigation is admittedly limited, the results are consistent with those already mentioned, and one must conclude that the onset of adolescence or adulthood in no way assures the availability of a complete combinatorial system.

Proportions and Correlations

We saw earlier that only 20 per cent of 90 college seniors were able to solve a "simple" problem correctly involving a balance arm and "weights" or masses. (see Figure 2–43) Since the problem was apparently not "simple" for four fifths of the students, something more than merely equalizing masses must be involved.

The concept of proportionality is precisely what seems to be the stumbling block in the balancing problem. Piaget refers to such a concept in terms of a "schema," meaning a coordinated set of mental structures that can be applied to a related problem, regardless of the context or content. Whatever terms are applied to the cognitive strategies involved, it will become apparent that where proportional thinking is required, even adolescents and adults encounter difficulty in metric or numerical solutions to problems.

The notion of correlation is closely related to that of proportion. Where one factor (variable) is related to another in an experimental situation, the correlation can be put in proportional form. The degree of correlation among factors is the knowledge outcome and the proportion is the "form," so to speak, which is used to reach the outcome. The association between proportion and correlation becomes clearer as we consider example experiments.

In *The Child's Conception of Geometry* (1948b), Piaget, Inhelder, and Szeminska suggest that the idea of proportions is first detected in children's perceptions of figures. One of the interview experiments involved a 1.5 centimeter by 3.0 centimeter rectangle. The child was asked to draw a rectangle that was "the same but larger" on another sheet of paper. Four broad stages were seen in the development of the notion of proportionality:

1. Children are unable even to draw what could be called a rectangle.
2. Drawing now looks like a rectangle but the ratio of side lengths is not maintained. The tendency is to exaggerate the length. There are no attempts to measure.
3. For the first time, children concentrate on both length and width and attempt to maintain the ratio, 1.5 to 3.0 or 1 to 2. For ratios different from 1 to 2, however, the children are unable to maintain the correct ratio.
4. Finally, the children are able to construct figures with many different ratios, such as 2/3 4/5 etc. This marks the beginning of understanding proportionality.

The proportion schema is actually a "second order" relation. A ratio is a "first order" relation and a proportion is a relation between ratios, or a relation of a relation. Inhelder and Piaget (1955) used a simple balance to study the development of the proportionality schema. Figure 2–48 shows the apparatus used in their interviews. Two unequal weights W and W' are balanced at unequal distances L and L' from the axis such that $W/W' = L'/L$. This inverse proportion is what could be called an equilibrium law. Until about the age of seven or eight years there are few indications that the subjects understand the relationship between weight and length. From this point to about the age of ten years, the children can coordinate equal weights with unequal distances and vice-versa, but coordination is not possible with unequal weights *and* unequal distances. This coordination is gradually achieved

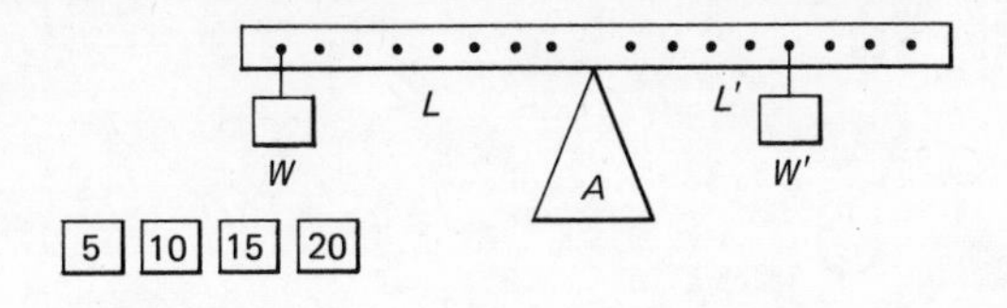

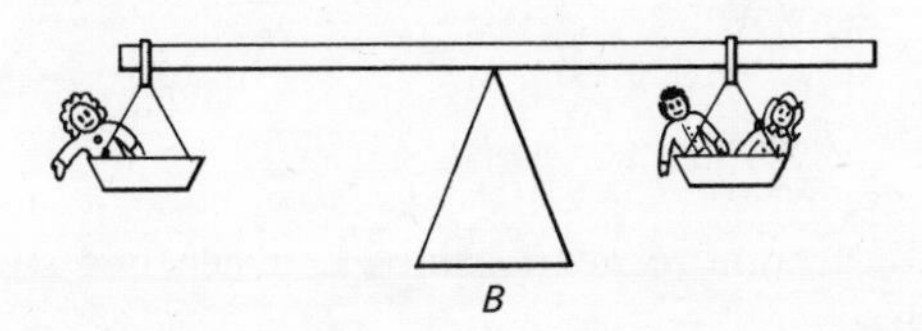

Figure 2–48. Balance apparatus used for studying proportions.

during the next 2 to 3 years so that by the age of twelve or thirteen years, Inhelder and Piaget found many children capable of discovering the equilibrium law of inverse proportionality.

A second experiment by Inhelder and Piaget (1955) approached the development of the proportions schema from a geometrical standpoint. Whereas the balance apparatus represents a dynamic physical equilibrium, their shadow projection apparatus in Figure 2–49 is of a static, geometrical nature. The law to be discovered is that the size of the shadow is directly proportional to ring diameter and inversely proportional to the distance from the candle. Children were asked to find two shadows of the same size, but using rings of unequal diameters. The investigators found a similar developmental stage with the proportions schema when compared with the balance apparatus. By the age of twelve or thirteen years, or a stage referred to as III-*A* by the authors, the children were able to calculate the correct proportional relationship between ring diameters and distances from the light source. During a final

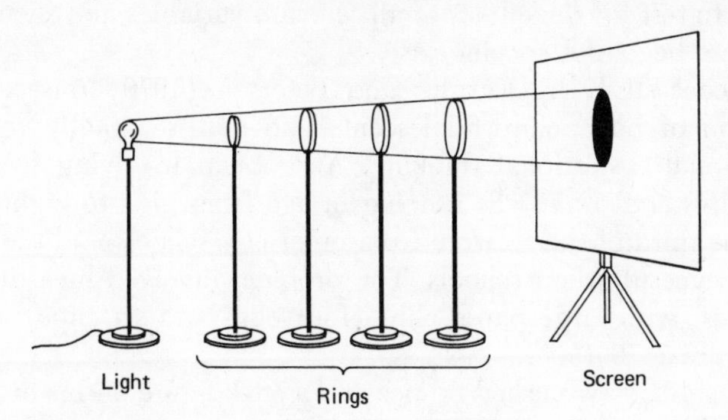

Figure 2–49. Projection of shadows apparatus used for proportions schema.

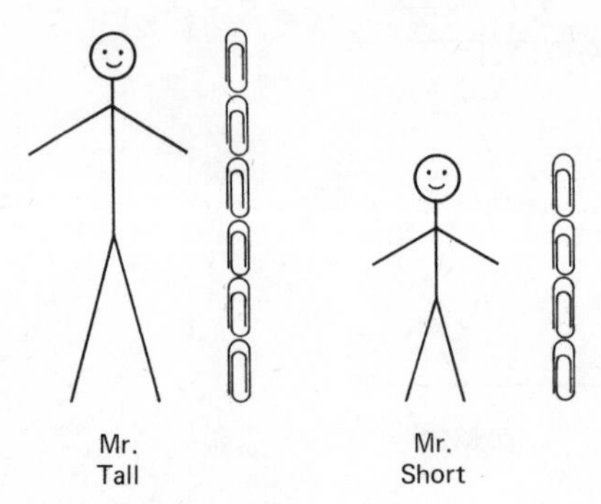

Figure 2–50. Demonstration of relationships between two stick figures and large paper clips.

stage, referred to as III-*B*, the children were able to generalize the law of inverse proportions to all possible combinations of ring diameter to distance from the light source, that is, the distances must have the same relation to each other as the rings.

In the balance experiment, as well as the shadows experiment, Inhelder and Piaget explain the attainment of generalizable laws in terms of a transformational system called the INRC group. Identity (*I*), negation (*N*), reciprocal (*R*), and correlativity (*C*) form a generalizable transformation system that allows the formal operational subject to form all possible combinations. Increasing a weight can be negated (*N*) by moving it closer to the axis or vice-versa. If weight and distance are increased on one arm of the balance, a reciprocal (*R*) action on the other arm can compensate and maintain equilibrium. This INRC group is analyzed in more detail later, but for now it should be seen as a very important part of Piaget's theory of formal operational thought because it allows the individual to consider all possible effects of changing variables in an experimental situation. This, of course, is essential if the experimenter is to test for the effects of all relevant variables and then be able to generalize beyond a specific case.

A more recent study by Karplus and Peterson (1970) raises the familiar question of how many adolescents and adults actually reach the stage of formal operational thinking. A problem involving length ratios was administered to 727 Ss ranging in age from nine to eighteen years. About one third of the Ss were from inner city schools and the remainder were from suburban schools. The problem involved measuring two "stick figures" with large paper clips (four clips and six clips) as a group. (see Figure 2–50).

Then, each student was asked to measure a stick figure on his or her paper with small paper clips made available to them. Three questions were asked:

1. How tall is this figure (Mr. Short), measured with small paper clips?
2. Predict the height of the large figure (Mr. Tall), as measured with small paper clips.
3. Explain how you figured out your prediction.

The responses from the 727 *Ss* were sorted into seven different categories:

1. No explanation or statement; "I can't explain."
2. *Intuition:* Data are not used in the explanation.
3. *Intuitive computation:* Data are used but in a haphazard and illogical way.
4. *Addition:* All data are used but differences rather than ratios are used.
5. *Scaling:* A change of scale is made by the subject while predicting and the scale inherent in the whole problem is ignored.
6. *Addition and scaling:* The excess height of Mr. Tall causes the subject to scale up the excess number of jumbo clips by a factor of two.
7. *Proportions:* True proportionality is used.

Table 2–19 shows the results of the study in percentages by group. Sixteen ambiguous responses were omitted. Four categories, 2, 4, 5 and 7, contain most of the 711 *Ss* in the study. Suburban students in grades 11–12 were far and away the most advanced group, with 80 per cent of the 153 *Ss* categorized as reflecting true proportionality. There are no great differences between the urban and suburban groups until the 8 to 10 grade span is reached. By then, 32 per cent of the suburban group is categorized as using proportional thinking, but only 5 per cent of the urban group is so categorized. This is a ratio of roughly 1 to 6. By the junior and senior year in high school, this ratio has jumped to 1 to 9.

TABLE 2–19. **Percentages of Students at Grade Levels in Each of the Seven Categories (Karplus and Peterson, 1970)**

	Grades (Number)						
Category	4–5 Suburban (116)	6 Suburban (82)	6 Urban (95)	8–10 Suburban (75)	8–10 Urban (123)	11–12 Suburban (153)	11–12 Urban (67)
1	2	1	5	0	0	3	1
2	31	36	41	4	16	3	15
3	9	5	12	11	21	1	12
4	16	26	20	25	30	10	31
5	33	26	16	24	25	3	20
6	4	0	3	4	3	0	12
7	5	6	3	32	5	80	9

By any reasonable estimate, the "paper clips" problem in proportions is fairly easy when compared to many problems found in science and mathematics textbooks in our high schools. The problem was presented in a concrete rather than a hypothetical way and no extraneous data were introduced. It would appear that the task was measuring proportional thinking at an early stage of development.

Inhelder and Piaget (1955) have noted that the structure of proportions always contains an element of compensation. To maintain the equality in a proportion (that is, $a/b = a'/b'$), as one factor is varied, a compensation must be instituted by varying one or more of the other factors. In real world experiments, where the relationships among variables can be rather complex, compensations become equally complex. For example, let us say that some outcome a has the following relationship, $a = \left(\frac{x}{y^2}\right)^{1/2} - 1$. Maintaining a constant outcome a, when one of the factors x or y varies, requires a rather complex compensation in the other factor. Formal operational thinking reflects an ability to identify all possible influential factors, quantify their degree of influence, and be able to maintain an equilibrium in the system through a series of compensations.

It was mentioned earlier that the notion or "schema" of correlation is closely related to that of proportion. The degree of correlation among various factors can be expressed quantitatively in proportional form. Inhelder and Piaget (1955) reported an experiment that was designed to assess Ss' abilities to calculate correlations between eye and hair color. In this test, each of 40 cards has a face drawn on it such that the eyes and hair are colored accordingly:

a = blue eyes and blond hair.

b = blue eyes and brown hair.

c = brown eyes and blond hair.

d = brown eyes and brown hair.

Each child is given a set number of cards and asked if the cards show a relationship between eye and hair color. If $a = b = c = d = 10$ or any other number, then the correlation (r) between eye and hair color is 0. Or if $a,b,c,d = 4,0,0,4$ then the correlation is 1. If blond hair is always associated with blue eyes and/or if brown hair is always associated with brown eyes, then the correlation between eye and hair color is 1. If $a,b,c,d = 4,2,2,4$ then the correlation is .333. Inhelder and Piaget (1955) examined the results of the interviews from the standpoint of propositional logic by using the following symbolism:

$p =$ presence of blue eyes.
$q =$ presence of blond hair.
$\bar{p} =$ absence of blue eyes.
$\bar{q} =$ absence of blond hair.

Combinations that are favorable to the correlation between eye and hair color are $(p \cdot q) \text{ v} (\bar{p} \cdot \bar{q})$, where "·" means "and" whereas "v" means "or." Thus, $(p \cdot q) \text{ v } (\bar{p} \cdot \bar{q})$ can be translated as *p and q or* not *p and* not *q*. Either combination is equally confirming of the correlation between blue eyes and blond hair *or* brown eyes and brown hair. The other two combinations, $(p \cdot \bar{q})$ or $(\bar{p} \cdot q)$ are confirming of no correlation between blue eyes and blond hair or brown eyes and brown hair. An elementary form of correlation can be expressed as

$$R = \frac{(a + d) - (b + c)}{a + d + b + c}.$$

The difference between the "positive" confirming cases and the "negative" confirming cases divided by the total number of cases will yield some number between -1 and 1. For example, the combination used before, 4, 2, 2, 4 (*a, b, c, d,* respectively) yields 1/3 or .333 . . . . Inhelder and Piaget (1955) reported that it was not until the age of fourteen to fifteen years that their Ss were able to understand the qualitative relationships involved in the correlation expressed just identified. A recognition of the reciprocal correspondence between $a(p \cdot q)$ and $d(\bar{p} \cdot \bar{q})$ enables the S in the "final" developmental stage, referred to as III-*B* and beginning at the age of fourteen to fifteen years, to make approximate calculations of correlation between eye and hair color. In symbolic logic terms, $p \cdot q = R(\bar{p} \cdot \bar{q})$ or $\bar{p} \cdot \bar{q} = R(p \cdot q)$.

The development of the ability to recognize reciprocal correspondence allows the S also to realize that $p \cdot \bar{q} = R(\bar{p} \cdot q)$ or $\bar{p} \cdot q = \text{R}(p \cdot \bar{q})$. Thus, the relationship between confirming cases, nonconfirming cases, and the total cases is internalized and correlation in a genuine sense becomes operational. Although exact calculations of correlation are not necessarily implied, the subject at this stage (III-*B*) can "see" the difference between the *a, b, c, d* sets 6, 2, 3, 4 and 4, 2, 2, 2 in terms of "strength" of the correlation between eye and hair color *or* that set 4, 3, 2, 6 is equal to 6, 2, 3, 4 in determining the correlation.

The notions or schemas of the proportion and correlation are related very closely to each other as well as to the combinatorial schema. Experimental situations in science require that the experimenter be able to use all three of these formal operational schema. Now that we have dealt briefly with these three "formal operational" schema, the following sections describe cognitive strategies referred to as "eliminating contradictions" and "identifying and controlling variables." These fundamental processes in experimental science are often introduced to children in elementary school. As is commonly the case, the cognitive requirements of the science curriculum very often exceed the cognitive abilities of the children.

Eliminating Experimental Contradictions

Many experimental situations confront the experimenter with data that are foreign or contradictory to beliefs held by the individual. Explanations for natural phenomena are often more complex than what might immediately be apparent on the surface of things. Inhelder and Piaget (1955) reported an experiment that was designed to identify the progression in children's abilities toward systematically eliminating apparent contradictions. A number of objects are presented to the child and she/he is asked to group them according to whether they think it (the object) will float or sink. Explanations for each classification are asked for, and then the child uses a bucket of water to test his predictions.

We saw earlier in the section on mass, weight, and density that density is a difficult concept, even for college seniors. Only about one third of those tested seem to have a useful concept of density. Figure 2–51 shows the type of question that usually results in a majority of adults giving incorrect answers.

The following information and questions that would accompany Figure 2–51 are[4]

1. A cylinder I contains three different liquids, each of a different color.
2. If more red liquid were added so that the cylinder contained about twice as much red liquid as blue liquid, how would the liquids appear (*a* or *b* or *c*?).

This type of item tests for a functional concept of density in relation to overall weight. Cylinder I suggests, by way of the relative amounts of each liquid, that "more" means "lower." If density has been operationalized in the mind of the subject, then the amount of substance will not alter its density. Three liquids with different densities (or specific gravity) will remain in the same relative position, as in Container I, for example, no matter how much of one or more of the liquids is added. The perceptual miscues in Figure 2–49 were sufficient to cause about two thirds of those adults (college seniors) responding to the item to choose either *b* or *c*. Density was not conserved when the amount of red liquid was increased significantly. Since density involves the coordination of weight with volume, it can be assumed that such coordination

[4] Thanks to J. Shymansky, friend and colleague, for this item.

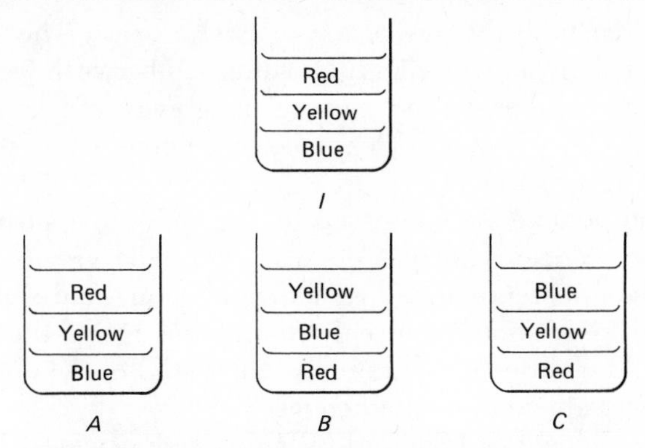

Figure 2–51. Density of liquids.

is not yet complete for those individuals who are unable to conserve density.

Returning to the Inhelder and Piaget experiments once again, the concept of density was central to the floating and sinking apparatus. Specific gravity can be defined as the ratio of the weight of a given volume of an object to the weight of an equal volume of water. A huge aircraft carrier floats simply because its weight per total volume is *less* than the weight of an equal volume of water. Conversely, a straight pin sinks because its weight per total volume is *more* than an equal amount of water. A third factor, surface tension of the water, can be a significant factor when the shape of the object approaches flatness. Aluminum foil, for example, has a specific gravity (and density) greater than that of water, but it will float on the surface because of the surface tension of the liquid. This means that three variables, weight, volume, and shape, must be considered when experimenting with floating and sinking to gain any consistent kind of explanation for the observable phenomena.

Inhelder and Piaget reported typical "stages" that they observed in the children's attempts to eliminate contradictions in the floating and sinking experiments. During stage 1 (up to the age of seven or eight years) children are satisfied with contradictiory explanations. They generally classify objects into two simple groups. Heavy objects will sink, regardless of size, and light objects will float. When confronted with certain obvious contradictions, the child in stage 1, according to Inhelder and Piaget (1955), attempts to solve the problem by offering an explanation that is in opposition to the reason just given.

During stage 2 (eight to ten years of age,) the children begin efforts to remove contradictions that they can now recognize, but they are only minimally successful. This should not be surprising since we saw in the sections on weight and volume that conservation is not attained by most children until the age of ten to twelve years. This prevents the child from making any meaningful relationship between weight and volume. Therefore, neither density nor specific gravity can be conceptualized and used to solve the apparent contradictions involved in the sinking and floating experiment.

After weight and volume are conserved by the child, it is possible to coordinate the two concepts so that density or specific gravity can be used to resolve the apparent contradictions with floating and sinking objects. Inhelder and Piaget noted that only toward the age of thirteen or fourteen years could children compare relative densities of the objects with water to eliminate apparent contradictions.

A recent study by Renner and Stafford (1972) raises questions about the thirteen to fourteen-year range that Inhelder and Piaget reported as the "operational" point involving the elimination of experimental contradictions. Five hundred and eighty-eight students in grades 7–12 were interviewed on six different tasks, one of which was similar to the floating and sinking experiment described earlier. The results showed that only about 10 per cent of that group could be called formal operational. A clear majority (67 per cent) was unable to eliminate successfully contradictions that emanate from weight, volume, and surface tension interrelationships.

The floating and sinking experiment is not atypical insofar as problem-solving investigations in science are concerned. There are few clear-cut, simple solutions. Most real problems require solutions that involve a number of variables, all contributing somewhat to the outcome. This fact leads us to the next section on identifying and controlling variables, probably the most important and difficult set of operations required in an experimental setting.

Identifying and Controlling Variables

The elementary school science curriculum project, *Science: A Process Approach* (SAPA), introduced the process of controlling variables for children in the upper elementary grades. Ideas for this curriculum were developed in the early 1960s and have been implemented, in one form or another, in many school systems in this country and abroad. Before the developers of this program decided that ten- or eleven-year-old children were capable of identifying and controlling variables, Inhelder and Piaget (1955) had reported results and conclusions from their research on the same processes. One experiment designed to assess children's abilities to separate variables involved the use of the materials shown in Figure 2–52. The rods vary by brass and nonbrass, round cross-section or square, large and small cross-section, length, and weight (added to the rod).

These five variables contribute to the degree of bending of each rod, and the subject must try to "separate out" these variables in order to determine the effect of each or of their combined effects. A large basin of water is placed directly under the rods and each subject is asked to determine whether a particular rod is flexible enough to reach the water. As with all experiments reported by Piaget, the subject's methods are observed and some type of proof is requested by the interviewer for the observations and explanations offered by the subject. Each rod is movable, to increase or decrease the rod's effective length, and the weight can be placed at any point on the rod.

Before the age of seven or eight years, Inhelder and Piaget reported that children took a random approach to the experimenting and that their explanations were "precausal" in nature. The ability to form logical class and relational properties has not developed by the age of seven or eight years to allow for a coherent, more systematic approach to the problem. During the next few years, as class and relational operations develop, children become more able to group the data (observations) and order them according to relative effects on the outcome (bending of the rod). First additive and then multiplicative classification schema become available. Factors (variables) are related first one-by-one, then two-by-two, then three-by-three, and so on as concrete classification ability develops. Simple correspondences between variables such as weight and the outcome (bending) or length and bending can be handled by successive trials, but as all of the variables are taken as an interrelated whole, that approach is inadequate. All variables must be separated, and each is varied such that all possible combinations are considered.

According to Inhelder and Piaget (1955), by the age of eleven or twelve years children begin to show signs of development of a more complete combinatorial system. Two characteristics of this "new" stage are *making hypotheses* about possible combinations and attempting

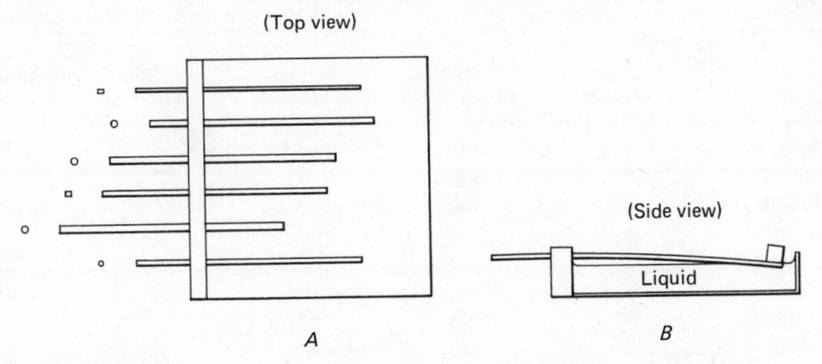

Figure 2–52. Apparatus for separating (identifying) variables.

proofs for them. The selection of *certain* pairs of rods rather than just *any* rods reflects the more sophisticated approach taken by the older subject. By the age of fourteen or fifteen years, Inhelder and Piaget suggest that most subjects are able to approach the experimental situation in a very systematic and exhaustive manner, reflecting what they refer to as formal operational thinking.

In a more recent study, Renner and Stafford (1972), using interviews derived from those reported by Inhelder and Piaget (1955), classified only 10 per cent of a sample of 588 children in grades 7–12 as "formal operational." In another study by Good (1973), 91 college seniors responded to a written question designed to assess the ability of an individual to analyze data collected from a science experiment (see Table 2–20). Each student was asked to determine from the data whether angle of inclination, weight, color, diameter, or *either* inclination or color affected the rolling time. Students were told to choose the *best* conclusion that was logically consistent with the data. By studying the data in Table 2–20, it can be seen that the two top data lines (*a* and *b*) show that hollow and solid cylinders have the same rolling time. Taking this into consideration, data lines *c* and *d* show that the diameter of the object has no effect on rolling time. If this strategy is followed for all of the data, it is found that color *or* angle of inclination affects the rolling time. Of the 91 college students who responded to the item, 34 or 37 per cent chose the correct response. Of course, guessing alone would give a 20 per cent correct result so the 37 per cent figure is in all likelihood a bit high. This assumption was supported by analyzing the students' *reasons* for their choices. Some admitted to guessing whereas others gave reasons that were not derived from the data. These results combined with those from the Renner and Stafford study (1972) certainly raise doubts about the extent to which adolescents and adults are capable of dealing logi-

TABLE 20. Data from "Rolling Cylinders" Experiment

Hollow or Solid	Diameter of Object	Weight of Object	Color of Object	Rolling Time	Angle of Inclination
a. hollow	2	50g	red	5	10
b. solid	2	50g	red	5	10
c. hollow	2	50g	red	5	10
d. solid	3	50g	red	5	10
e. hollow	2	35g	red	5	10
f. solid	3	35g	blue	3	10
g. hollow	3	50g	red	3	20
h. solid	2	35g	red	5	10

cally with experimental situations involving more than two or three variables.

In another experiment involving four apparent variables, Inhelder and Piaget (1955) assessed children's ability to exclude noninfluential variables after they have been isolated. Pendulum apparatus (see Figure 1–5 in Part One) included four apparent variables: length of the string, weight of the object hung on the string, height of the release point, and the degree of "push" applied to the object. Only one variable, length of string, has a significant effect on the frequency of swings. The previous experiment with flexible rods included five variables, each one influential in causing bending. Inhelder and Piaget reported results that were very similar to those described in the flexible rods experiment. Classification and seriation abilities develop into a complete combinatorial system so that all possible correspondences are considered by the subject after the age of fourteen or fifteen years. The transition from simple to multiple classes and relations, and finally to a complete combinatorial system involving propositional logic, allows the experimenter to consider hypothetical correspondences between variables and outcome(s), even though it may not be possible to actually demonstrate the correspondence in a concrete way. The characteristics of the system of propositional logic, which allow for such a formal or abstract approach to problems, are now considered.

Propositional Logic

Propositional thought is characterized by a complete combinatorial system that allows for separation and control of variables and a generalized hypothetical-deductive aproach to solving problems. Prepropositional thought is limited by an incomplete combinatorial system and only concrete reality is a part of the problem-solving domain. A comparison of the combinatorial system that is characteristic of prepropositional thought with that of propositional thought should help to clarify the distinctions.

Let us say that some factor F is being investigated in association with an outcome O. The child who has an "incomplete" combinatorial system will be limited to the following combinations:

1. $F \times O$ (O occurs with F present),
2. $F \times O'$ (O does not occur with F present),
3. $F' \times O$ (O occurs when F is not present),
4. $F' \times O'$ (O does not occur when F is not present).

Inhelder and Piaget (1955) have noted that 1 to 1 class multiplication is involved in such an approach and the subject discovers the combinations in a concrete manner of experimentation. Sometimes, the nine to eleven- or twelve-year-old child will also include 2 by 2 class multiplication, but all combinations are not exhausted.

Propositional thought presupposes the ability to make all possible combinations even though it may not be possible to actually verify each one experimentally. Each possibility represents a hypothetical grouping. Whereas before we saw that simple class groupings produced only four combinations of the experimental variable (F) with an outcome (O), a system of propositional logic will produce 16 combinations. The symbols p and q are used here to represent any variable (p) and an associated outcome (q). When a line appears above p and/or q, it will mean that one or the other or both do not appear in the particular combination. Propositional conjunction (and) and disjunction (or) signs $\cdot$ and v, respectively, are used to "connect" the possible combinations. If it helps to make the following symbolism more concrete, the reader might want to let p represent a specific experimental variable such as a particular chemical involved (or at least present) in a chemical reaction (the outcome q). The problem is to determine if p is a causal factor in relation to q. The following 16 combinations are possible using a propositional combinatorial system:

1. $p \cdot q$ (q occurs with p present).
2. $p \cdot \bar{q}$ (q does not occur with p present).
3. $\bar{p} \cdot q$ (p does not occur with q present).
4. $\bar{p} \cdot \bar{q}$ (q does not occur and p is not present).
5. $(p \cdot q) \vee (p \cdot \bar{q})$ (1 above *or* 2 above).
6. $(p \cdot q) \vee (\bar{p} \cdot q)$ (1 above *or* 3 above).
7. $(p \cdot q) \vee (\bar{p} \cdot \bar{q})$ (1 above *or* 4 above).
8. $(p \cdot \bar{q}) \vee (\bar{p} \cdot q)$ (2 above *or* 3 above).
9. $(p \cdot \bar{q}) \vee (\bar{p} \cdot \bar{q})$ (2 above *or* 4 above).
10. $(\bar{p} \cdot q) \vee (\bar{p} \cdot \bar{q})$ (3 above *or* 4 above).
11. $(p \cdot q) \vee (p \cdot \bar{q}) \vee (\bar{p} \cdot q)$ (1 above *or* 2 above or 3 above).
12. $(p \cdot q) \vee (p \cdot \bar{q}) \vee (\bar{p} \cdot \bar{q})$ (1 above *or* 2 above *or* 4 above).
13. $(p \cdot q) \vee (\bar{p} \cdot q) \vee (\bar{p} \cdot \bar{q})$ (1 above *or* 3 above *or* 4 above).
14. $(p \cdot \bar{q}) \vee (\bar{p} \cdot q) \vee (\bar{p} \cdot \bar{q})$ (2 above *or* 3 above *or* 4 above).
15. $(p \cdot q) \vee (p \cdot \bar{q}) \vee (\bar{p} \cdot q) \vee (\bar{p} \cdot \bar{q})$ (1 *or* 2 *or* 3 *or* 4).

---

16. $\overline{(p \cdot q) \vee (p \cdot \bar{q}) \vee (\bar{p} \cdot q) \vee (\bar{p} \cdot \bar{q})}$ (neither 1 nor 2 nor 3 nor 4).

It should be emphasized that these combinations are only *theoretical* possibilities. In a real experimental setting, the investigator will, in all likelihood, not use each and every possible test. For example, if the chemical p is present and the outcome q is not present $(p \cdot \bar{q})$, then the investigator might conclude that p is not a causal factor. Conversely, if q is present and p is not present $(\bar{p} \cdot q)$, a similar conclusion might be reached. When both tests are conducted $(p \cdot \bar{q})$, $(\bar{p} \cdot q)$, the conclusion that p is not causal would seem to be justified. The operation of implication $(p \supset q)$ means that the presence of the independent variable p implies the presence of the outcome (dependent variable) q. Another way of showing implication is the following: $(p \supset q) = (p \cdot q) \vee (\bar{p} \cdot \bar{q}) \vee (\bar{p} \cdot \bar{q})$. Table 2–21 shows the five different "types" of implication that are theoretically possible in an experimental setting where the effects of a particular variable p are being investigated relative to an outcome q. Implication operations are very common in science investigations, since finding out how something "works" usually involves explanations citing causal relationships, for example, "*If* you lengthen the string, *then* the pendulum swings more slowly (rate); *If* more mass is added to this side of the balance, *then* more mass must be added to the other side; *If* an egg hatches into a chick, *then* we conclude it was fertilized. Bynum et al. (1972) reported results of a study in which the use of all 16 binary propositional operations was tested. They concluded that at least six of the formal operations were seldom, if ever, used, including two implication operations, inverse of implication (nonimplication) $(p \cdot \bar{q})$ and inverse of converse implication $(\bar{p} \cdot q)$. Since $(p \cdot \bar{q})$ is a part of converse implication, $(p \cdot q) \vee (p \cdot \bar{q}) \vee (\bar{p} \cdot \bar{q})$, and $(\bar{p} \cdot q)$ is a part of implication $(p \cdot q) \vee (\bar{p} \cdot q) \vee (\bar{p} \cdot \bar{q})$, it is not difficult to see why Bynum et al. found that students never use these two implication operations.

To realize the need for controlling variables, the investigator must be able to "see" all of the possible relationships between variables and outcome(s). If simple conjunction $(p \cdot q)$ or conjunctive negation

TABLE 2–21. Implications Operations in the Set of Sixteen Possible Binary Operations

Name of Operation	Symbolism
1. Implication:	$(p \cdot q) \vee (\bar{p} \cdot q) \vee (\bar{p} \cdot \bar{q})$
2. Inverse of implication (nonimplication):	$(p \cdot \bar{q})$
3. Converse implication:	$(p \cdot q) \vee (p \cdot \bar{q}) \vee (\bar{p} \cdot \bar{q})$
4. Inverse of converse implication:	$(\bar{p} \cdot q)$
5. Reciprocal implication:	$(p \cdot q) \vee (\bar{p} \cdot \bar{q})$

$(\bar{p} \cdot \bar{q})$ is used to determine the "cause-effect" relationship between variable p and outcome q, it is very likely that the investigator will wrongly conclude a causal relationship. For example, in the flexible rods experiment conducted by Inhelder and Piaget (1955), there were five variables; length, weight, type of metal, diameter or rod, and shape or cross-section, all of which contributed to the outcome (flexibility). The process of verifying the causal relationships between variables and outcome presupposes a workable combinatorial system. It might be useful to identify some possible experiments within the flexible rods investigation (see Figure 2–52) and describe them in symbolic logic terms. The following terms are used:

p = long rod, $\bar{p}$ = short rod.

q = heavy rod, $\bar{q}$ = light rod.

r = brass rod, $\bar{r}$ = nonbrass rod.

s = large diameter rod, $\bar{s}$ = small diameter rod.

t = round cross-section, $\bar{t}$ = square cross-section.

x = large bending effect, $\bar{x}$ = small bending effect.

With only one variable (p) and an outcome (q), we saw that there are 16 combinations that are theoretically possible. With two variables and an outcome, 256 combinations are theoretically possible. With five variables; length, weight, material, diameter, and cross-section shape and an outcome, flexibility, there are obviously a very large number of possible combinations, so a strategy must be devised to reduce effectively the number of variables. The strategy, of course, is to select two rods so that all characteristics except one are the same. In symbolic terms $(p \cdot q \cdot r \cdot s \cdot t \cdot x) \vee (p \cdot q \cdot r \cdot \bar{s} \cdot t \cdot x)$, which means that the two rods differ only by diameter. In effect, this procedure reduces the effective number of variables to one, diameter. If the outcome, bending, is the same in both cases, then it would be reasonable to conclude that diameter, by itself, does not affect bending. If the outcome is different $(p \cdot q \cdot r \cdot s \cdot t \cdot \bar{x}) \vee (p \cdot q \cdot r \cdot \bar{s} \cdot t \cdot x)$, then it would be reasonable to conclude that small diameter rods bend more than large diameter rods, all other things being equal. The situation then is reduced to $(\bar{s} \supset x) \vee (s \supset \bar{x})$, if small diameter rods then large bending *or* if large diameter rods then small bending. Reducing the experiment to a simple implication makes the five variable experiment much easier to cope with. The verification schema of "all other things being equal" is characteristic of the individual who is capable of formal or abstract thinking.

This schema can be used again and again to "exhaust" the possible factors affecting the outcome, x.

We now look into how the 16 binary operations are interrelated into a system. Piaget has consistently emphasized the interrelatedness of the cognitive system and the following section describes how his proposed INRC group forms a totally interrelated system of transformations.

The INRC Group

The 16 binary operations are logical relations between variables (p) and outcomes (q). Piaget has devised a logical model, the INRC group, which describes how the formal thinker interrelates the 16 binary relations into a dynamic system. This does not mean that conscious effort is needed to accomplish the interrelationships, but merely that the individual has developed the necessary cognitive structures. Identity (I), negation (N), reciprocity (R), and correlativity (C) represent the four "transformations" that can transform any of the binary operations ($p \vee q$), ($\bar{p} \vee \bar{q}$), ($p \cdot \bar{q}$), and so on into any other.

Identity (I). As the name implies, this transformation changes nothing. For example, $I(p \vee q) = (p \vee q)$, or $I(p \cdot \bar{q}) = (p \cdot \bar{q})$.

Negation (N). This transformation is important because it transforms *all* aspects of relations between variables and outcomes. For example, $N(p \vee q) = (\bar{p} \cdot \bar{q})$. The assertions p and q are changed from positive to negative *and* the logical connective or (v) is changed to and (·). Other examples should help to further clarify the function of N:

$$N(p \cdot q) = (\bar{p} \vee \bar{q})$$
$$N(\bar{p} \vee \bar{q}) = (p \cdot q)$$
$$N(p \cdot \bar{q}) = (\bar{p} \vee q)$$
$$N(p > q) = (\bar{q} > \bar{p})$$
$$N(p \cdot q) \vee (\bar{p} \cdot q) = (\bar{p} \vee \bar{q}) \cdot (p \vee \bar{q})$$

Reciprocal (R). This transformation changes the assertions p and q from positive to negative and vice-versa, but leaves the conjunctions and disjunctions unchanged. $R(p \cdot q) = (\bar{p} \cdot \bar{q})$ or $R(\bar{p} \vee q) = (p \vee \bar{q})$.

Correlative (C). This transformation changes the conjunctions to disjunctions and vice-versa but leaves the assertions p and q unchanged. $C(p \cdot q) = (p \vee q)$ or $C(\bar{p} \vee \bar{q}) = (\bar{p} \cdot \bar{q})$.

These four transformations, I, N, R, C are interrelated such that $I = NRC$, $N = IRC = RC$, $R = INC = NC$, and $C = INR = NR$. We can show this to be true in the following way:

$$\text{Identity, } I = NRC$$
$$I(p \vee q) = (p \vee q)$$
$$N(p \vee q) = (\bar{p} \cdot \bar{q})$$
$$R(\bar{p} \cdot \bar{q}) = (p \cdot q)$$
$$C(p \cdot q) = (p \vee q)$$
$$\text{thus } I = NRC$$
$$\text{Negation, } N = RC$$
$$N(p \vee q) = (\bar{p} \cdot \bar{q})$$
$$R(p \vee q) = (\bar{p} \vee \bar{q})$$
$$C(\bar{p} \vee \bar{q}) = (\bar{p} \cdot \bar{q})$$
$$\text{thus } N = RC$$
$$\text{Reciprocal, } R = NC$$
$$R(p \vee q) = (\bar{p} \vee \bar{q})$$
$$N(p \vee q) = (\bar{p} \cdot \bar{q})$$
$$C(\bar{p} \cdot \bar{q}) = (\bar{p} \vee \bar{q})$$
$$\text{thus } R = NC$$
$$\text{Correlative, } C = NR$$
$$C(p \vee q) = (p \cdot q)$$
$$N(p \vee q) = (\bar{p} \cdot \bar{q})$$
$$R(\bar{p} \cdot \bar{q}) = (p \cdot q)$$
$$\text{thus, } C = NR$$

All other relations, in addition to $(p \vee q)$, can also be used to show the complete interrelatedness of the INRC group. Inhelder and Piaget (1955) used a balance apparatus (see Figure 2–46 in the earlier section on proportions and correlations) to illustrate the use of the INRC group in the solution of the problem. An understanding of the system of negations (N) and reciprocals (R) involved in weight and distance (from the fulcrum) changes leads to an understanding of the proportional nature of equilibrium in the balance. The identity operation, I, was defined as $(p \cdot q)$, where p is a fixed increase in weight added to one of the pans of the balance and q is a fixed increase in the distance of the same pan from the fulcrum. Therefore, $\bar{p}$ represents a fixed decrease in weight on the same pan and $\bar{q}$ represents a fixed decrease in the distance of that pan from the fulcrum. Likewise, p', q', $\bar{p}'$, and $\bar{q}'$ represent similar operations on the other pan of the balance. The INRC group transforms the $(p \cdot q)$ relations in the following ways:

$I(p \cdot q) =$ Increasing the weight and the distance on one pan.

$N(p \cdot q) = (\bar{p} \vee \bar{q}) = (p \cdot \bar{q}) \vee (\bar{p} \cdot q) \vee (\bar{p} \cdot \bar{q}) =$ Increasing the weight while reducing the distance *or* reducing the weight while increasing the distance *or* reducing the weight while reducing the distance on one pan.

$R(p \cdot q) = (p' \cdot q') =$ Compensating I by increasing weight and distance on the other pan.

$C(p \cdot q) = (\bar{p}' \vee \bar{q}') = (p' \cdot \bar{q}') \vee (\bar{p}' \cdot q') \vee (\bar{p}' \cdot \bar{q}') =$ Canceling R in the same way N cancels I.

Negation in a physical sense means the annulling of a particular operation by actually eliminating it, whereas a reciprocal action opposes the operation in question (such as adding weight to one of the balance pans) without actually annulling it. When the equilibrium in a balance is destroyed by adding weight *and* moving the pan away from the fulcrum $(p \cdot q)$, the equilibrium can be restored by removing the weight and moving the pan back to its original position (negation). The equilibrium can also be restored by adding weight to the other pan and moving it a similar distance away from the fulcrum (reciprocal, $p' \cdot q'$). There are, of course, other ways to restore an equilibrium in a balance that has been destroyed by changes p (adding weight) and q (adding distance from the fulcrum). Adding enough weight on the other pan (p') *or* increasing the distance of the other pan from the fulcrum (q') could also restore the equilibrium. In quantitative terms, $p \times q$ must equal $p' \times q'$ to have a state of equilibrium in the balance. An understanding of the system of negations and reciprocals is directly related to the following relations:

$$\frac{p \cdot q}{p \cdot q} = \frac{p \vee q}{p \vee q} \text{ or } \frac{Ix}{Rx} = \frac{Cx}{Nx} \; (x = p \cdot q)$$

An operational understanding of the interrelatedness of the INRC group allows the individual to develop logical proportions and thus maintain an equilibrium in a mechanical or physical system such as a movable pan balance.

The perceptive reader may have noticed a discrepancy between the purely logical INRC group and the physical transformations involved in "real" experimental situations such as in the balance system. Negation in the balance system means to annul or "undo" some change in the system, but in the logical group, $N(p \cdot q)$ is equivalent to $(\bar{p} \vee \bar{q})$. Since $(\bar{p} \vee \bar{q})$ means $(p \cdot \bar{q}) \vee (\bar{p} \cdot q) \vee (\bar{p} \cdot \bar{q})$, it leaves open the possibility that only q or only p is negated. This makes little sense if negation is to have any real meaning in a physical setting. When an operation such as adding weight and increasing distance $(p \cdot q)$ is negated, all of it, not just part of it is negated. It is the reciprocal transformation R that actually seems to negate $(p \cdot q)$ in a physical sense $[R(p \cdot q) = (\bar{p} \cdot \bar{q})]$. This discrepancy would seem to justify questioning the complete applicability of the logical INRC group as offered by Piaget in describing the actual operations in a physical setting. We saw earlier that of the five implication-type binary propositions (out of the total of 16), only three seem to have any real application in a physical setting. Thus, the theoretical model based on the 16 binary propositions and the INRC group developed by Piaget seems to predict reality only in an approximate way. This does not make it any less valuable, however. The theory is truly comprehensive in nature and is based heavily on research findings. The discrepancies or variations referred to between theory and so-called "reality" are a common phenomena between all theories and experimental confirmations. The "fit" is approximate, ragged, and necessarily "imperfect." Theories serve as guides and change as our empirical probings become more and more accurate.

It was mentioned in the earlier section on probability that Piaget identified eight formal structured operational schemas. They include combinations, proportions, acceleration, mechanical equilibrium (balance), probability, correlation, multiplicative compensations (flexible rods), and advanced forms of conservation. Inhelder and Piaget (1955) note that the organization of the formally structured schemas grows out of the INRC group–propositional logic system, which as they assert, *is* the structure of formal thought. Piaget attributed three characteristics that the eight schemas have in common: (1) They are generally applicable to a great many problem-solving situations; (2) They are more deduced or abstracted from existing mental structures themselves than from discovery in objects; (3) They are related to the propositional logic-INRC group system.

1. Schema of combinatorial operations.

   This schema allows for a systematic method in solving problems involving combinations and permutations. Although the combinatorial schema is obviously closely related to the set of 16 binary propositional operators, it is not a part of the set.

2. Schema of proportions.

   This schema does derive from the INRC group-propositional operations system. Proportions of the form $\frac{a}{x} = \frac{b}{y}$ merely relate two ratios. This general form is extremely useful in the quantitative solutions to many and varied problems in science.

3. Coordination of two reference systems. Included in this schema are relativity of motion or acceleration. All motion studies where two moving objects are a part of the system of study require the use of this schema.

4. Schema of mechanical equilibrium.

   This schema requires the coordination of negation (N) and reciprocity (R). The equilibrium of a mechanical system such as a balance is maintained through the application of this schema.

5. Schema of probability.

   This schema is linked closely to the combinations schema. Chance concepts form the basis of many important fields of science.

6. Schema of correlation.

   Correlation, probability, and proportions are all very closely interrelated. Trying to determine the relationship between variables obviously involves the system of propositional logic.

7. Schema of multiplicative compensations.

   This schema and proportions are closely related. In the volume of solids, for example, as the form is altered, multiple compensations must be implemented to "conserve" the overall volume.

8. Schema of "advanced" forms of conservation.

   "Advanced" refers to forms that are very difficult to verify empirically. Momentum conservation is an example of such an advanced form.

What all of these eight schemas have in common is that they are rooted in the propositional logic–INRC group system. This means they are linked together as a system and can be applied individually or in various combinations to solve problems involving proportions, combinations, probability, and so on.

Summary of Development at an Abstract Level

Formal thinking is required in science or any other knowledge-generating endeavor such that explanations, theories, and so on must be consistent with the data collected. As new data are generated, these explanations of physical causality in science must be revised to make for a better "fit" of the data. Abstract or formal thinking as defined by Inhelder and Piaget (1955) is centered around the binary system of propositional logic and the INRC group that transforms each of the operations (for example, $p \cdot q$) into a completely interrelated, operational system. The formal schemas of combinations, proportions, and others that grow out of this formal operational system are seen as components throughout the whole of science. As the mental structures necessary for abstract thinking develop, the individual becomes more capable of reasoning with propositions and hypotheses, eventually to the exclusion of any corresponding physical reality. What is *possible* now dominates over what is *real.* Formal or abstract thought allows for the isolation and systematic control of real variables as well as for the identification of *potential* relationships.

In summarizing any developmental characteristics, one runs the risk of suggesting that static kinds of stages actually occur. In the summary of concrete thinking, this danger was identified and it is again raised here. There are no specific beginnings or endings in cognitive development, merely a continuous process of mental adaptation. Stages have been developed as an artificial means to communicate some very complex findings on children's thinking. It is not their thinking that develops in stages, but our attempts at explaining, in summary form, a tremendously complex area of study. It is hoped that these precautionary statements will place in perspective the following attempts at summarizing what has been called formal thought development. Table 2–22 lists some char-

TABLE 2–22. **Characteristics of Abstract as Compared to Concrete Thought**

Concrete Operations	Formal or Abstract Operations
1. Classifications and serial ordering are employed in a step-by-step fashion, without relating all of the links into a system.	1. Reasoning now begins with propositions and hypotheses. A system of what is hypothetically possible is structured and followed by empirical verification.
2. Well-mixed variables in an investigation cannot be separated.	2. Factors or variables in an experimental setting are now seen as distinct and can be separated from one another.
3. Logical multiplication of factors is limited to 1 by 1 or 2 by 2 correspondences but a total *n* by *n* system is not yet available.	3. A complete combinatorial system is available so that *all* combinations of factors (*n* by *n*) can be exhaustively tested.
4. Reversibility in operations is limited to negation *or* reciprocity but they are not interrelated as a system.	4. Reversibility is advanced by the interrelationship of negation and reciprocity, resulting in the ability to maintain a dynamic equilibrium in a system involving many factors.
5. Experimental variability is seen as a result of multiple causes and no systematic efforts are made to isolate and control factors or variables.	5. Variables can now be controlled systematically.
6. Chance or probability ideas are influenced by previous results, yielding a kind of "gambler's fallacy."	6. The notion of probability becomes operational and widely applicable.
7. The notion of correlations is incomplete, and errors occur when anything other than simple relationships are involved.	7. Correlations are no longer limited to 1 or 1/2, but are, at least qualitatively, also applicable in such situations as 1/6, 2/7, 3/11, and so on.
8. Proportions are not as yet available except in the simplest sense.	8. Proportions are now much more widely applicable in solving problems.
9. Experimental contradictions cannot be pursued in a systematic manner.	9. Because of the complete combinatorial system, experimental contradictions can now be isolated and solved.
10. Conservation is limited to those physical qualities that are easily tested.	10. Conservation is no longer limited to the immediate, testable environment.
11. Thinking is a derivative of the child's own actions on concrete reality.	11. Coordination of reference systems is possible and acceleration is worked out.

Concrete Operations	Formal or Abstract Operations
12. "Models" represent the concrete reality that seems to be offered rather than abstract possibilities worked out by others.	12. In an experimental situation involving many variables, logical pairs are sorted out and tested.
	13. The utilization of a proof based on "all other things being equal" is now pursued.
	14. The binary system of propositional logic based on the operations of conjunction, disjunction, implication, and others is operable.
	15. The INRC group transforms the binary system into a fully operational, interrelated system.

acteristics of concrete thought in one column and those for formal or abstract thought in the other. Although Piaget has suggested that formal thought structures develop by the age of fourteen or fifteen years, many other sources raise questions about the percentage of adolescents and adults who actually develop such capabilities.

The preceding characteristics are really centered around the development of propositional logic operations that are governed or interrelated by the INRC group. Since these have already been described in some detail, further elaboration is not provided. Piaget has noted that the real "power" of propositional logic is the combinatorial system that allows reality to be analyzed in terms of possible hypotheses. Thought can now be directed from what is possible to what is real.

A complete combinatorial system allows for the systematic identification, separation, and control of variables in an experimental setting. Once the "complete" combinatorial system is developed, the subject can isolate apparent contradictions in data through the systematic control of variables and combinations thereof. Theory can be developed or comprehended at the logical or formal level because of the ability to analyze possibilities in terms of internal consistency. "Thinking about thinking" is a characteristic of the formal operational person. Thoughts that are generated can be analyzed in terms of "fit" within a hypothetical system as well as applicability to a real system. Abstract ideas generated by mathematicians require no reality at all for validation, merely internal consistency within specified rules of logic. The abstract nature of the work of mathematicians and scientists is perhaps the "temporary ultimate" model of the nature of logical thought development. The human mind surely reflects its structure and temporary capabilities through the actions of human beings.

references

ALLEN, THOMAS. "The Effect of Optical Illusions on the Conservation of Length." An unpublished study done in a graduate course in the Department of Science Education, Florida State University, Tallahassee, 1972.

ANNETT, MARIAN. "The Classification of Instances of Four Common Class Concepts by Children and Adults." *British Journal of Educational Psychology,* **29**, 223–36, 1959.

BITTNER, A. C., AND M. SHINEDLING. "A Methodological Investigation of Piaget's Concept of Conservation of Substance." *Genetic Psychology Monographs,* **77**, 135–65, 1968.

BLAIR HOOD, H. "An Experimental Study of Piaget's Theory of the Development of Number in Children." *British Journal of Psychology,* **53**, 273–86, 1962.

BYNUM, TERRELL, JAMES THOMAS, AND LAWRENCE WEITZ. "Truth-Functional Logic in Formal Operational Thinking." *Developmental Psychology,* **7**, 129–32, 1972.

COHEN, J., AND C. HANSEL. "The Idea of Independence." *British Journal of Psychology,* **46**, 178–90, 1955.

DAVIES, C. M. "Development of the Probability Concept in Children." *Child Development,* **36**, 779–88, 1965.

DELACY, ELIZABETH. "Some Problems Associated with a Paper and Pencil Test of Conservation of Length." *Child Development,* **38**, 1967.

DODWELL, P. C. "Children's Understanding of Spatial Concepts." *The Canadian Journal of Psychology,* **17**, 141–61, 1963.

ELKIND, DAVID. "Children's Discovery of the Conservation of Mass, Weight, and Volume: Piaget Replication Study II." *Journal of Genetic Psychology,* **98**, 219–27, 1961.

EMERY, JOYCE. "The Status of Certain Probability Concepts and Combinatorial Abilities of High School Biology Students and the Effect of Genetics Instruction on these Cognitive Characteristics." Unpublished Ph.D. dissertation, Department of Science Education, Florida State University, Tallahassee, 1973.

FLAVELL, JOHN. *The Developmental Psychology of Jean Piaget.* New York: Van Nostrand Reinhold Company, 1963.

FLOOD, PAUL. "Identification, Diagnosis, and Implications of Matter Change Concepts Expressed by High School and College Students." Prospectus of a dissertation submitted to the Department of Science Education, Florida State University, Tallahassee, 1973.

Good, Ronald G. "Coordination of 2- and 3-Dimensional Perspective in Grades K–4." An unpublished study conducted at the Florida State University Developmental Research School, Tallahassee, 1972.

———. An unpublished study reporting the results with college seniors, of a 16-item instrument designed to assess selected cognitive characteristics of adolescents and adults. Department of Science Education, Florida State University, Tallahassee, 1973.

———, and Charles Matthews et al. "Cognitive Development Characteristics of Children in Grades 1–5." A part of a comprehensive study presented at the 46th Annual Meeting of the *National Association for Research in Science Teaching,* Detroit, 1973.

———, and James Shymansky. "Perimeter–Area Relations and Geometrical Limits Concept in Children, K–5." An unpublished study conducted at the Florida State University Developmental Research School, Tallahassee, 1973.

Gratch, G. "The Development of the Expectation of the Nonindependence of Random Events in Children." *Child Development,* **30**, 217–27, 1959.

Green, Paul, M. Gringras, and M. Neff. "A Study of Conservation of Water Level." An unpublished study conducted by graduate students in a graduate course in science education. Department of Science Education, Florida State University, Tallahassee, 1973.

Inhelder, Barbël, and Jean Piaget. *De la logique de l'enfant a la logique de l'adolescent.* Paris: Presses Universitaires de France, 1955; translated by A. Parsons and S. Seagrin, *The Growth of Logical Thinking From Childhood to Adolescence.* New York: Basic Books, Inc., 1958.

———, and ———. *La genese des structures logiques elementaires.* Neuchatel: Delachaux and Niestle, 1959; translated by E. Lunzer and D. Papert, *The Early Growth of Logic in the Child.* London: Routledge and Kegan Paul, 1964.

Jarvik, M. E. "Probability Learning and a Negative Recency Effect in the Serial Anticipation of Alternative Symbols." *Journal of Experimental Psychology,* **41**, 291–97, 1951.

Karplus, Robert, and Rita Peterson. "Intellectual Development Beyond Elementary School II: Ratio, A Survey." Science Curriculum Improvement Study, University of California, Berkeley, 1970.

Kofsky, Ellin. "A Scalogram Study of Classificatory Development." *Child Development,* **37** (1), 1966.

Laurendeau, Monique, and Adrien Pinard. *Causal Thinking in the Child.* New York: International Universities Press, 1962.

Lovell, Kenneth. "Some Studies Involving Spatial Ideas" in *Piagetian Cognitive-Development Research and Mathematical Education.* The National Council of Teachers of Mathematics, 1971.

Lovell, K., and I. Butterworth. "Abilities Underlying the Understanding of Proportionality." *Mathematics Teaching,* **37**, 1966.

———, and E. Ogilvie. "The Growth of the Concept of Volume in Junior School Children." *Journal of Child Psychology and Psychiatry,* **2**, 118–26, 1961.

———, D. Healey, and A. Rowland. "Growth of Some Geometrical Concepts." *Child Development,* **33,** 1962.

Lunzer, E. "Problems of Formal Reasoning in Test Situations." *Monographs for the Society of Research in Child Development,* **30,** (2), 1965.

———. "Formal Reasoning." In *Development in Learning,* Vol. 2, edited by Lunzer and Morris. London: Staple Press, 1968.

McNair, Wes, and Mike Slack. A replication of the "mountains task" on perspective done in a graduate science education course at Florida State University, Tallahassee, 1972.

Minichiello, M., and J. Goodnow. "Effect of an Action Cue on Conservation of Amount." *Psychonomic Science,* **16,** 200–01, 1969.

Mori, Ichio, and Nikom Tadang. "The Effect of Abnormal Speed Motion Picture Films on a Child's Spatio-Temporal Recognition, Part II: On the Comparison of Synchronous and Isochronous Events." *Science Education,* **57,** 325–30, 1973.

Piaget, Jean, and Barbël Inhelder. *Le développement des quantites chez l'enfant.* Neuchâtel: Delachaux and Niestlé, 1941.

———, and ———. *La genèse de l' idée de hasard chez l' enfant.* Paris: Presses Universitaires de France, 1951.

Piaget, Jean. *Le développment de la notion de temps chez l' enfant.* Paris: Presses Universitaires de France, 1946a; translated by A. Pomerans, *The Child's Conception of Time.* London: Routledge and Kegan Paul, 1969, Basic Books, Inc., Publishers, New York.

———. *Les notions de movement et de vitesse chez l' enfant.* Paris: Presses Universitaires de France, 1946b; translated by G. Holloway and M. Mackenzie, *The Child's Conception of Movement and Speed.* London: Routledge and Kegan Paul, 1970.

———. *La causalité physique chez l' enfant.* Paris: Librairie F. Alcan, 1927; translated by M. Gabain, *The Child's Conception of Physical Causality.* Totowa, N.J. Littlefield, Adams, 1960.

———, and ———. *La representation de l' espace chez l' enfant.* Paris: Presses Universitaires de France, 1948a; translated by F. Langdon and J. Lunzer, *The Child's Conception of Space.* London: Routledge and Kegan Paul, 1956.

———, and Alina Szeminska. *La genese du nombre chez l' enfant.* Neuchâtel: Delachaux and Niestlé, 1941; translated by C. G. attegno and F. Hodgson, *The Child's Conception of Number.* London: Routledge and Kegan Paul, 1952.

———, and A. Morf. "Les préinferences perceptives et leurs relations avec les schèmes sensori-moteurs et opératoires." in *Etudes d' epistemology genetic,* Vol. VI, *Logique et perception,* 1958.

———, Barbël Inhelder, and Alina Szeminska. *La géometrie spontanée de l' enfant.* Paris: Presses Universitaires de France, 1948b; translated by E. Lunzer, *The Child's Conception of Geometry.* London: Routledge and Kegan Paul, 1960.

Phillips, Darrell G. "The Development of the Concept of Displacement Volume: A Hierarchical Model and Its Partial Testing Under Two Methods of Presentation." *Journal of Research in Science Teaching,* **8,** 9–19, 1971.

RAVEN, RONALD. "The Development of the Concept of Momentum in Primary School Children." *Journal of Research in Science Teaching,* **5,** 216–23, 1967–68.

———. "A Multivariate Analysis of Task Dimensions Related to Science Concept Learning Difficulties in Primary School Children." *Journal of Research in Science Teaching,* **9,** 207–212, 1972a.

———. "The Development of the Concept of Acceleration in Elementary School Children." *Journal of Research in Science Teaching,* **9,** 201–206, 1972b.

RENNER, JOHN W., AND DONALD STAFFORD. *Teaching Science in the Secondary School.* New York: Harper & Row, Publishers, Inc., 1972.

ROBINSON, LUCILE T. "An Exploratory Study of the Utilization of Spatial Concepts by Children." *Dissertation Abstracts,* **31,** 1631-A, October, 1970.

ROSS, B. M., AND N. LEVY. "Patterned Predictions of Chance Events by Children and Adults." *Psychological Reports,* **4,** 87–121, 1958.

SMOCK, C. D., AND G. BELOVICZ. "Understanding of Concepts of Probability Theory by Junior High School Children." Final Report, Wisconsin R & D Center for Cognitive Learning, 1968.

TOWLER, JOHN O., AND GRAYSON WHEATLEY. "Conservation Concepts in College Students: A Replication and Critique." *Journal of Genetic Psychology,* **118,** 265–70, 1971.

part three

A THEORY OF COGNITIVE DEVELOPMENT

objectives

It is intended that this section will help the reader to

1. Recognize some important terms used by Piaget in his theory of cognitive development.
2. Identify major stages of cognitive development.

introduction

Theory has been largely avoided to this point in favor of an emphasis on descriptions of experimental interviews and results. It is usually Piaget's theory that attracts the greatest amount of attention from most psychologists and educators, whereas the rich store of data that led to the gradual formulation and validation of the theory is somehow forgotten or relegated to a position of relative unimportance. To determine the validity of any theory, however, data are required. In Piaget's case, large amounts of data were gathered in support of a theory of cognitive development. Other data collected by independent researchers support the sequence of development reported by Piaget, but significant questions have been raised about the chronological ages that generally are associated with the developmental sequence. Particularly, the development of abstract or formal structures and schema seem to be far less widespread among adolescents and adults than suggested by Piaget. These variations do not, however, have any significant negating effect on his theory.

Piaget's background as a biologist is evident in his overall explanations on how intelligence develops to enable human beings to create knowledge. Intellectual development is merely a special form of overall biological development. Genetic determination of development is seen by Piaget as playing a relatively small part in intellectual progress. Apart from our biological limitations, such as "seeing" only certain wavelengths of light and hearing only certain frequencies of sounds, heredity is subordinated to the nature of our interaction with the environment in cognitive development considerations. The nature of this interaction is an *adaptation* composed of two active components, *assimilation* and *accommodation.* Adaptation in the general physiological sense is the process whereby the developing organism adjusts to the environment to survive and continue its own kind. *Assimilation* is that part of adaptation in which the developing organism changes the environment to fit its characteristics. If food represents a part of the environment, the chewing and digestive processes of the body change the foodstuffs to adapt them to the benefit of the growth and maintenance of the organism. There are also corresponding adjustments made by the organism to *accommodate* to the food. This dynamic adaptation process involving an equilibrium between *assimilation* and *accommodation* has been referred to by Piaget as *equilibration.* All acts of intelligence involve a dynamic equilibrium between the two adaptive processes, *assimilation* and *accommodation.*

At birth, the human organism is capable of very few intelligent acts. The mental *structures* necessary for such behavior have not been constructed. It is Piaget's thesis that such structures do not just "automatically" develop as physiological growth occurs, but must be constructed, as it were, by active intervention with the environment. This active process, known as *equilibration,* simultaneously *assimilates* a "piece" of the environment and *accommodates* mental structures to better understand or adapt to the piece of reality. "Grasping a ring" by the three-month-old baby is something that must be developed laboriously through successive structuring. The mental *schema* or set of structures necessary for grasping a ring must be developed through the active process of *equilibration.* The cognitive structures become more adaptive to the reality of the ring and, of course, more generalized so that future "ringlike" objects can be assimilated more readily. All intellectual functioning involves assimilation of the external environment to existing mental structures while at the same time altering the structures to better adapt to that "piece" of reality. Two examples of cognitive functioning, where there is an imbalance between assimilation and accommodation, are *dreamlike play* and *imitation.* Dreamlike play is a small subset of the entire world of play and does not refer to the "self-structured" activity of children that adults usually observe. The play of building a special kind of house for the first time with blocks is very different from dreamlike activity where *reality* is molded or altered to fit the whims of the child. Dreamlike play strongly favors assimilation with very little accommodation on the part of the child. In other words, there is little attempt to understand or accomplish anything new. Instead, reality is changed to fit into existing cognitive structures. At the other extreme, *imitation* involves changing cognitive structures or *accommodating* without any corresponding reality. The mental equilibrium is now strongly in the direction of accommodation. Whenever dreamlike play or "pure" imitation is the primary activity of the organism, intelligent adaptation is minimal. The "normal" state of affairs in·cognitive development involves a proper equilibrium between assimilation and accommodation. *Equilibration* assures adaptation of the organism to its environment through the continual development of structures, to integrated structures or schemas, to integrated schemas that form an interrelated cognitive system. The means for this equilibrating process are the child's *actions* on objects in the immediate environment. Actual physical activity is central to Piaget's theory of cognitive development and what follows is a relatively brief look at four developmental periods that have come to be associated with his theory. These periods should not be construed or interpreted as anything except a continual developmental process. Mental structures and schemas develop continuously; there are no abrupt divisions or periods inside the bony shell. It is for convenience of communication that stages or periods or other summaries are invented.

sensory–motor period zero to two years

During the late 1920s and early 1930s Piaget and his wife had two daughters (1925, 1927) and a son (1931). All three of the Piaget children became the "objects" of intensive study by the Piagets. Their observations became the basis for two books: *The Origins of Intelligence in the Child* (original, 1936 and translation, 1952) and *The Construction of Reality in the Child* (original, 1936 and translation, 1954). Most of the material for this section on sensory-motor development is drawn from these two sources.

From birth onward, the child interacts with the immediate environment and cognitive structures develop through the process of *equilibration.* In this sense, the child is constantly in a *transition* stage as new structures continue to develop into *schemas* such as grasping, kicking, and reaching. These schemas develop as an interrelated system through the child's continual *equilibrative* process between *assimilation* and *accommodation.*

During the first few weeks of life the child depends largely on reflexes for its continued existence, the most important of which is the sucking reflex. Experience in using the sucking reflex causes modifications of this inherited schema such that it becomes more sophisticated. It is exercised even between feedings because, as Piaget theorizes, there is a natural tendency to exercise schema to ensure their adequacy and elaboration. Thus, the sucking schema is exercised by the baby between feedings and there is a resultant elaboration of this schema to better adapt to the required function and, ultimately, to increase the odds for survival. This tendency to seek stimulation for developing schemas is termed "functional assimilation" by Piaget and is central to his theory of intellectual development.

After the first few weeks, the infant develops behavior patterns that go beyond the preliminary reflex reactions. Motor coordination develops that allows the child to place objects in the mouth and the first signs of curiosity in nearby, surrounding objects are observed. It is apparent that before about the age of four months, the child sees objects such as rattles, rings, and other "transient" things as existing only as long as they are within the child's sensory apparatus. By about the age of one year, the child definitely engages in intentional acts such as moving an object in its path to reach another desired object.

During the second year of life, the child searches for new things merely for the sake of finding out about new situations. This exploratory behavior extends developing schemas farther and farther into larger segments of the child's world. New means of exploration are discovered as the available schemas are exercised, and with the advent of crawling and walking the regions available for exploration are greatly expanded. Objects that disappear from sight are searched for where they were last seen and are understood as having permanence even though they may disappear for long periods of time. Nearer the end of the second year the child begins to show evidence that it can mentally represent objects that are no longer present. This kind of "symbolic thought" marks the beginning of genuine thinking in a kind of problem-solving sense. This symbolic function signals the beginning of representational thought, which is generally referred to as the "preoperational period."

preoperational period (two to seven years)

With the onset of the symbolic function, representational thought evolves and allows for a vastly expanded set of schemas. Separate events begin to be linked together, although not systematically, and thinking can reflect a genuine searching for solutions when the child is confronted with a problem. The thought is still very egocentric, however, and "other" points of view are not within the child's frame of reference. There is no reason to justify ideas since only one point of view seems possible. Perceptions generally center on one particular feature of an object, thus neglecting other potentially important aspects. This lack of ability to compensate for changes in a system evokes "nonconserving" responses from the preoperational child. Successive transformations cannot be assimilated.

Another important characteristic of preoperational thought is the lack of "reversibility." The child is unable to reverse the order of events in thought and arrive once again at the beginning point. This feature of thinking from two to seven years of age is perhaps emphasized most by Piaget as the major obstacle to real "operational" thinking. A genuine mental *operation* is reversible (by definition) so the preoperational child does not actually engage in operational thought. It is important, of course, to remember that a gradual evolution of mental schemas during the preoperational period occurs and the child is constantly progressing toward "operational" thought. The six-year-old is vastly more advanced than the three-year-old even though both may be included in Piaget's preoperational period of development. The reader should not "lose" this concept of continuous development in the neatly categorized periods and stages, that tend to oversimplify the real nature of cognitive development.

concrete operational period (seven to eleven years)

Since so much time has already been devoted to this period in Part Two, little elaboration seems necessary, except to familiarize the reader with some of the more common terms associated with the theory as compared to the experiments.

An operation refers to an internalized act, representational in nature and linked to other operations into a system. Piaget has divided operations during the concrete operational period into two main groups; logical and infralogical. The logical operations such as establishing one-to-one correspondence, adding, and subtracting are not tied to any specific quantities, space, time, or the like as are the infralogical operations. Piaget has likened infralogical operations to the logical-mathematical structures used by the mathematician in classes and relations. For Piaget, the cognitive structures are reflected by these mathematical structures. Concrete operations require objects that can be grouped or ordered. These concrete groupings involve reversibility by either *inversion* (adding or removing objects) or *reciprocity* (compensating for a change) but not by a synthesis of the two as in the more advanced "formal" operations. The groupings are only simple or multiple classifications rather than a complete *n*-by-*n* combinatorial system.

formal operational period (eleven years on)

When the adolescent accomplishes a complete *n*-by-*n* combinatorial system, problems of a concrete as well as an abstract nature can be solved. Problems involving many variables can be attacked systematically through separation and control techniques such that *all* possibilities are examined or at least considered in a theoretical sense. Formal operations have been described in some detail in Part Two, and so a further elaboration here seems unnecessary. The logical model formulated by Piaget is centered around the 16 binary operations and the INRC group and is really reflected in the hypothetical-deductive logic system of the mathematician.

Piaget's theory has had enormous impact on developmental psychology and so the lack of detail here in reference to the theory should not be interpreted as a "slight" to the importance of the theoretical aspects of his work. It is merely that our purposes in this book require that theory be de-emphasized in favor of the applied, experimental side of things.

part four

SCIENCE CONCEPTS AND PROCESSES: COGNITIVE REQUIREMENTS

objectives

It is intended that this section will help the reader to

1. Differentiate between science as content and science as process.
2. Identify example science concepts commonly found in first grade textbooks.
3. Compare the conceptual requirements of science concepts in first-grade texts with the developmental abilities of six- and seven-year-old children.
4. Analyze first-grade science texts in terms of conceptual requirements.
5. Repeat objectives 2, 3, and 4 for third- and fifth-grade concepts.
6. Differentiate between concrete and abstract science concepts.
7. Analyze science texts using other criteria such as *amount* of information, and nature of demonstrations suggested.
8. Identify science processes as emphasized in some science curricula such as Science: A Process Approach.
9. Describe example science activities that can be used to allow the various science processes to function at kindergarten, first-, third-, and fifth-grade levels.
10. Compare the developmental conceptual abilities of children with various science processes expected of them in kindergarten, first-, third-, and fifth-grade levels.

introduction

The first three parts of this book have been devoted to describing the cognitive characteristics of children. In particular, Part Two emphasized the development of science-related concepts and processes such as classification, seriation, number, space, time, and motion. The nature of logical or abstract thinking was described in terms of the appearance of such schemas as probability, combinations, and proportions and as a system of propositional logic. All of the information was intended to provide the reader with a basis for making rational decisions about curriculum and instruction. Planning and implementing a physical fitness program for children obviously would require a thorough understanding of their physical characteristics. It is very likely that a wide range of physical abilities would be possessed by any given group of children, even those with similar chronological ages. Planning and implementing a science program also requires a thorough understanding of children's characteristics, but in this case, cognitive rather than physical characteristics. Of course, the goals that one chooses will be a major determining factor in any curriculum development or teaching efforts, but a consideration of such issues will be postponed until Part Five: The Role of the Teacher.

The nature and structure of science have often been presented as a set of facts, concepts, and principles that are "given" to us by someone else. More recently, attempts have been made to unify blocks or strands of content into larger "conceptual schemes" such as atomic-molecular nature of matter, relativity, populations, matter-energy relations, ecosystems, genetic nature of life, and electric and magnetic interaction. Whether the "content" is grouped together into "conceptual schemes" or presented in a more piecemeal fashion, this approach to science is certainly the most common approach, particularly at the postelementary levels.

One curriculum development project in particular, *Science: A Process Approach,* interpreted science in a very different way. Content, per se, was not the primary concern of the developers. Instead, science was defined and presented as a series of processes. The developers chose to define science, at least in a broad sense, as a "process approach." Processes of observing, classifying, measuring, and the like supposedly led to more advanced processes such as making hypotheses, controlling variables, and finally, experimenting. This emphasis on "the processes of science" was seen as the true essence of science by the developers of this curriculum project. Processes, of course, require a medium to occur and the medium was drawn from the traditional disciplines or groupings of physical sciences, biological sciences, and mathematics. Other curriculum development projects such as *Science Curriculum Improvement Study* (SCIS), *Elementary Science Study* (ESS), and others also tended to emphasize *process* over *product.* One in particular, SCIS, claimed to attend to Piaget's work in developmental psychology as an important base in developing science ideas for children.

Both of these interpretations of science, content and process, are used in attempting to determine the cognitive requirements of selected portions of science curricula and textbooks. All educators seem to agree that "appropriate" learning experiences should be provided for children, regardless of their grade level. Something is cognitively appropriate when the individual is capable of making sense of the situation, whether it is centered around content or process. Something can be "figured out" or understood only to the extent that there exists a corresponding operation, schema, set of schemas, or the like that ensure a stable equilibrium between assimilation and accommodation. Piaget's theory of mental adaptation and development is firmly based on the conclusion that the developing child must physically interact with objects to ensure his or her optimum growth toward logical or formal operations. In a sense, the child must become an integral part of the events that he or she actively pursues.

The reader might want to review the summaries of concrete and abstract thinking in Part Two before continuing into the following sections on the cognitive requirements of selected science concepts and principles.

science concepts and children's thinking

Most of the science curriculum materials available to teachers and children in schools are centered around facts and concepts. Sometimes the authors group the content into larger "packages" and call them "conceptual schemes." Since there are a great many different textbook series and other curriculum materials available for use in the schools, it is not feasible to look at all examples in terms of children's thinking. Some examples of science concepts intended for various grade levels are described and interpreted in light of the information in Part Two on cognitive characteristics. At the elementary school level, a wide range of science concepts with an even wider difficulty range appear in the various textbook series available for local and state adoption. This be comes increasingly clearer as examples for grades 1, 3, and 5 are presented and compared with the general cognitive characteristics of the particular grade level. Since a 2- to 3-year range in developmental thinking ability is not uncommon at any grade level, it is a risky business to determine the appropriateness of selected science concepts. In a great many instances, however, there is such a large discrepancy between the general cognitive characteristics of a given group of children and the cognitive requirements of the science content that "fine-line" analyses are unnecessary.

Grade One: Science Concepts and Children's Thinking

Children's thinking at this age (six to seven years) is generally "precausal" in nature. That is, things happen because objects are seen as having a special kind of "aliveness," especially the stars, moon, sun, clouds, and certain other inanimate objects. Also, children at this age commonly attribute causality to God or some other "maker" (artificialism). Contradictions often arise in their explanations without any awareness or concern on the part of the child. Children in first grade are usually led by their perceptions rather than their thinking as is evidenced by the numerous conservation studies. A consistent, thinking approach, *even with objects,* is generally beyond children at this age. Beginning classification and seriation abilities are developing, but cannot be applied well in all but the simplest problem-solving situations. Logical classes and subclasses are still beyond most children as are such number-related abilities as singular class, multiple classes, and ordinal–cardinal coordination. Some topological relations are functional but others such as circular order, knot tying, continuity, and the like are developing, but prefunctional. Projective and Euclidean concepts are not as yet functional for most first grade children, although conservation of length is not uncommon. Perspective is still egocentric and time and motion concepts are still prefunctional. Objects represent themselves, generally, rather than representing "models" of more abstract or inaccessible phenomena.

In general, it can be said that the typical six- or seven-year-old child in first grade is still tied largely to his or her perceptions and thinking is mostly precausal in nature. Symbols, even pictures, of portions of the child's environment are often confusing, unless what is represented very nearly resembles what the child has experienced, in all dimensions. Size in pictures or diagrams of insects, for example, can be very confusing, even though the child may have interacted with different kinds of insects. First grade children have difficulty in size translations and when pictures in textbooks are incomplete in some way children are often unable to "see" what the author(s) intended. For instance, if only part of an object is shown in the representation, the first grade child may be unable to "complete" the object mentally, even though it is fairly familiar. Thinking, at this level, is facilitated primarily through "actions on objects," rather than through words and pictures or diagrams.

Science Concepts Commonly Found in Grade One *Textbooks*

Science textbooks, and, in fact, almost all printed matter, are usually divided into "units" and "subunits" or overall concepts and subconcepts. One of the common units or overall concepts found in science textbooks intended for first grade children is on *weather.* Since first grade children are usually just beginning to read, most of these books consist of pictures, sketches, diagrams, and the like, which the teacher uses to "discuss" related concepts. *Models* are occasionally used to *represent* some phenomena and demonstrations are suggested, but, because of the nature of the content, an "actions on objects" approach by the children is not really feasible. Some of the following concepts found in first-grade textbooks are related to weather characteristics and changes:

1. Sunshine, temperature, moisture, and wind are elements of weather.
2. Wind blows from different directions.
3. Air is all around us.
4. Wind is air in motion.
5. Wind has force.
6. Water is always evaporating.
7. When water evaporates, it goes into the air and becomes invisible.
8. Clouds and precipitation result from the cycle of evaporation and condensation.
9. The cycle of evaporation and condensation is a result of heat exchange.
10. The weather cycle is related to the water cycle.
11. Changes in the state of matter are determined by a loss or gain of heat.
12. The amount of surface area of water is related to evaporation.
13. Evaporation, condensation, cloud formation, and precipitation occur in a cycle.

For children who generally

1. Exhibit precausal thinking (animism, artificialism, and the like).
2. Do not recognize contradictions, even in their own statements.
3. Are usually led by their perceptions in making decisions.
4. Classify and order objects only in a very simple manner.
5. Understand only one perspective, theirs.
6. Cannot "imagine" things unless they are concrete in nature *and* a part of their experience.
7. Do not link events together as a whole.
8. Do not relate cause-and-effect events except in the very simplest sense.

9. Have no stable conception of direction.
10. Have difficulty in tying their shoes.

is seems a bit ridiculous to expect these same first-grade children to gain any type of genuine understanding of such concepts as those in the previous list. Only the most superficial, fragile kind of knowledge can be gained when the gap is so wide between what is to be learned and the mental structures available for the job.

Another commonly taught science area or conceptual scheme in first grade is that of the *earth, moon, sun system.* Some of the concepts typically found in science textbooks (particularly the teacher's guides) include the following:

1. The sun is the star nearest the Earth.
2. The sun's energy provides heat and light.
3. Day and night are caused by the rotation of the Earth.
4. Planets, including the Earth, revolve around the sun.
5. The sizes of the planets vary as do their relative distances from the sun.
6. The moon reflects sunlight.
7. Phases of the moon are a result of the relative position of the Earth, moon, and sun.
8. Constellations are groups of stars that are many millions of miles from the Earth.
9. The sun's light "blocks out" the star's light.
10. The universe is constantly changing.

Relative size, position, motion, time, and distance concepts are involved in one way or another with most of these ten concepts. Classification ability is also involved to the extent that a sun can also be a star and vice-versa, and the Earth is also a planet and all are different in many ways. Much more than simple classification ability is required to make sense of such a varied "system" called a solar system. Children at the age of six or seven years have great difficulty conceptualizing the relative meaning of a town or city without having expectations that "Earth" or "world" can truly be assimilated and accommodated (internalized) in any sensible way. This again reflects the primitive classification ability of first-graders. Rotation of the earth assumes a developmental level *at least* 2 to 3 years beyond most first-grade children. Perspective, relative position, and motion must be internalized and coordinated to gain any genuine understanding of the relative sizes, positions, and movements of planets and other parts in our solar system. It is only by the age of nine or ten years that children can *begin* to make sense of such concepts, and then it is only a beginning, where one can see an occasional glimmer of understanding. Circular motion is more difficult to understand and time and motion studies reported in Part Two point up the 2- to 3-year lag behind linear motion.

The gap between conceptual requirements of typical units on the solar system, and the range of conceptual abilities of first-grade children, is very great indeed. For children who are still at the point where they believe the sun, moon, or stars "hide from them at will" or that the sun does not shine at night because "it needs time to rest," concepts such as those just described can have little meaning. Also, the learning method is necessarily mostly talking, listening, or watching. Pictures in books and teacher's words provide little opportunity for the six- or seven-year-old to develop mental structures through their own actions on concrete objects. Developing a coordinated perspective, even for real objects in their immediate environment occurs through countless actions and not until the age of ten or eleven years are most children able to perceive a system of simple objects from a variety of viewpoints. The far-removed, rather abstract notion of a solar system or even a related and visible Earth-sun-moon system is most certainly more difficult to "operationalize" than a three-dimensional "mountains task" described in the earlier section on projective geometry.

Many other conceptual schemes often appear in first-grade science textbooks and other curriculum materials that require a developmental level usually not available to most six- and seven-year-old children. The reader is encouraged to analyze available textbooks and other materials in terms of the concepts intended for first grade children. In addition to "Concrete Thinking" in Part Two and the related summary, Table 4–1, which follows, shows a type of checklist that predicts the existence, in part, of developmental abilities of children six and seven years old. A truly comprehensive checklist would be so long as to be of little use, so the list in Table 4–1 contains characteristics that have a fairly good likelihood of appearing, in some form, in science curriculum materials. Most of the same characteristics can be used for other grades, although some additions and deletions will be necessary. The predicted presence or absence of the cognitive characteristics is indicated by a mark under either the *yes* or *no* column. This will become increasingly difficult to do for grades 3 and 5 since development is gradual and few things are so neatly categorized as a simple yes or no.

Having a simple two-category system, yes or no, is most certainly an oversimplification. Cognitive development is a continuous process and, as such, a child is always in a transition stage, going from a particular

TABLE 4–1. **A Checklist for Use in Analyzing Science Concepts Available to *First Grade* (Six to Seven Years) Children**

Description or Requirement	Likelihood of Availability Yes	No
1. Actions on objects approach required.	x	
2. Objects represent only themselves (no models).	x	
3. Simple classification only.	x	
4. Simple seriation only.	x	
5. Simple counting.	x	
6. Decisions governed by perceptions.	x	
7. Measure lengths consistently.		x
8. Estimate distances.		x
9. Causal thinking (cause and effect).		x
10. More than one perspective (not egocentric).		x
11. Objects can be interrelated as a whole or a system.		x
12. Stable coordinate system.		x
13. Functional conception of area or volume.		x
14. Continuous conception of time (duration, sequence).		x
15. Consistent conception or measurement of motion.		x
16. Weight conservation.		x
17. Recognize contradictions.		x
18. Imagine "unseeable" objects (e.g., atoms, molecules).		x
19. Imagine events not experienced.		x
20. Identify variables.		x
21. Control variables.		x
22. Identify all combinations.		x
23. Deal with chance or probability.		x
24. Make correlations.		x
25. Make ratios or proportions.		x
26. Use deductive logic in "proofs."		x

point toward another. Using Table 4–1, without having a corresponding understanding of development as "continuous transition," is likely to be highly misleading and result in unjustified conclusions and overgeneralizations. Such summaries can be useful if these potential dangers are used to temper the tendency to think and act in terms of absolutes such as "yes" or "no."

Grade Three: Science Concepts and Children's Thinking

Eight- and nine-year-old children have made considerable advances since first grade, but cognitively they are still far away from logical or abstract thinking. Fewer children in third grade are so strongly tied to precausal thinking notions of artificialism, animism, and the like, although most still use some of these notions in their explanations about physical causality. Third grade children are more likely to become aware of contradictions in their observations or reasons for things happening. The range of abilities among children is fairly wide though, and even the more advanced ones can deal only with very simple observations and reasons related to objects and events in their immediate environment. They are still quite dependent on their perceptions, but there are some indications, such as the appearance of reversibility in some operations, conservation of number in most children, and conservation of continuous and discontinuous substance, that a "thinking approach" is becoming

TABLE 4–2. A Checklist for Use in Analyzing Science Concepts Available to *Third Grade* (Eight to Nine Years) Children

Description or Requirement	Likelihood of Availability		
	Yes	Some	No
1. Actions on objects approach required.	x		
2. Objects represent only themselves (no models).	x		
3. Multiple classes and stable classification criteria.		x	
4. Hierarchical classes.		x	
5. Seriation by length or width.	x		
6. Simple number operations.	x		
7. Decisions governed by perceptions.		x	
8. Accurate measurement (linear only).		x	
9. Estimate distances.		x	
10. Causal thinking.		x	
11. Coordinated perspective.		x	
12. Objects can be interrelated as a whole.		x	
13. Functional conception of area.		x	
14. Functional conception of volume.			x
15. Generalized reference system.			x
16. Simple conception of time.	x		
17. Functional conception or measurement of motion.			x
18. Weight conservation.		x	
19. Recognize contradictions.		x	
20. Imagine "unseeable" objects. (e.g., atoms)			x
21. Imagine events not experienced.			x
22. Identify variables.			x
23. Control variables.			x
24. Identify all combinations.			x
25. Chance or probability.			x
26. Correlations.			x
27. Ratios or proportions.			x
28. Use deductive logic in proofs.			x

more common. Classification is now firmly established in that intensive and extensive properties are coordinated for simple classes and criteria for forming certain groups or classes are stable. Hierarchical classification, however, is still not generally possible. That is, logical groupings, subgroupings, sub-subgroupings, and the like cannot be consistently constructed, except through rather extensive trial-and-error attempts. Measurement of length and distance are now far more functional, but area and volume conservation and measurement are generally undeveloped. The ability to view objects or systems of objects from various perspectives is far more advanced, when compared to the six- and seven-year-old, but mistakes are often made for all but rather simple systems. A generalized reference system that allows the child to locate points in space is still at least 2 to 3 years from the stage of being functional in the sense of solving even simple problems. A stable concept of time is fairly common, although accurate time measurement where linear and/or circular motion are involved is not common. Table 4–2 summarizes the likelihood of children in third grade having a functional grasp of selected abilities, many of which appeared in the previous table.

Keeping in mind the 2- to 3-year range of developmental levels among children in third grade, the general characteristics listed in Table 4–2 can be used to analyze science concepts found in available curriculum materials. A wider range of science concepts can be found in third-grade textbooks as compared to first-grade texts. An inspection of eight widely adopted textbook series showed the following concepts presented in one form or another:

1. How living things get air, food, water, and heat.
2. The effects of air and sun on regions of the earth.
3. The role of the sun's heat, light, and gravity in the solar system.
4. Matter–energy relationships in reactions involving heat.
5. Relationships between forces and movement.
6. Formation of rocks and soil.
7. Role of friction in simple machines.
8. Physical states of matter.
9. Sounds as vibrations.
10. Electrical forces.
11. Refraction and reflection of light.
12. Cells in living things.
13. Chemical properties and the role of electrons.

14. Chemical changes.
15. Atomic energy.
16. Causes of weather.
17. Interactive forces (gravity) in solar system.
18. Photosynthesis in plants.

The level of difficulty and abstractness can, of course, vary a great deal within the domain of any specified concept or conceptual scheme. Some concepts have been selected and described on the following pages, with a corresponding description of the primary mode of presentation. The reader is encouraged to make some "best guesstimates" as to the appropriateness of the science content intended for third-grade children (eight to nine years).

Concept 1: Where Does Energy Come From?[1]

A sketch depicts a body of water, some shoreline, clouds, and the sun (see Figure 4–1).*

Words accompanying the sketch inform the child that

a. Heat from the sun changes the water to an invisible gas.
b. The gas goes into the air.
c. Clouds are formed and they go over the earth and it rains.

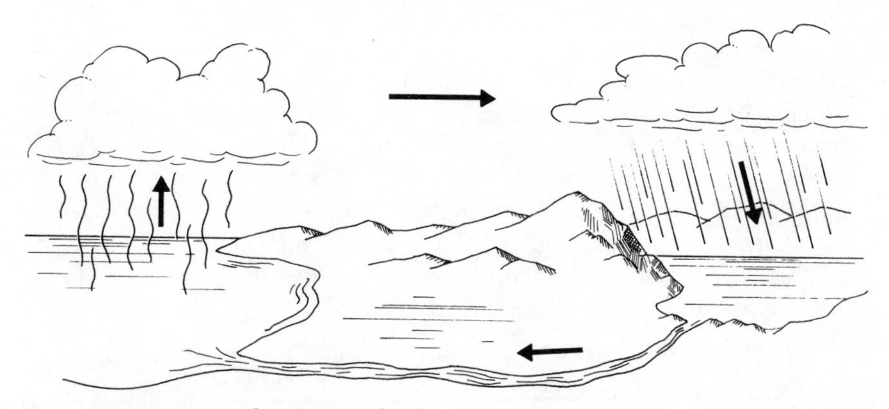

Figure 4–1. Sketch in a third-grade textbook depicting the sun as the source of energy.

[1] Illustrations on pp. 167 and 170 are reproduced by permission of Harcourt Brace Jovanovich, Inc., from CONCEPTS IN SCIENCE, Teacher's Edition, Level 3 (Green) by Brandwein, Cooper, Blackwood, Hone and Frazer; copyright © 1972 by Harcourt Brace Jovanovich, Inc.

The teacher is encouraged to demonstrate evaporation and to discuss the role of the sun's energy in the evaporation-condensation cycle.

Concept 2: The Nature of Sound

The teacher's guide for this concept explains that children will learn that sounds

1. Are vibrations carried by air, water, and the like.
2. Are waves that move in all directions.
3. Can be rapid vibrations (high-pitch) or slower vibrations (low-pitch).
4. Take time to travel.

The primary mode of instruction is with the teacher leading discussions about sketches in the students' books. Vibrations are emphasized with drums, tuning forks, water waves, and the like. Large and small drums, string instruments, and flower pots are used to illustrate the relative speed of vibrations and the corresponding pitch. One sketch shows a drum being hit and concentric rings around the drum, apparently moving out to a person's ear, which then "hears" the invisible waves.

Concept 3: Interaction of Force, Motion, and Friction.[2]

A force is defined for the children as a "push or pull that changes the motion of an object." Questions in the textbook ask each child about changes in direction of objects such as leaves, toy cars, and sleds, all intended to focus the children's thinking on conservation of motion. Sketches on the various pages depict (for those who can comprehend) various objects, either moving or still. Gravity is introduced with a sketch similar to that in Figure 4–2 and then defined as a force that pulls everything "down" toward the center of the earth.

Down is defined as "toward the center of the earth" and *up* is defined as "away from the center." The following statements about force and gravity are found in the child's book:

1. When a helicopter's blades push it up into the air, the force of the spinning blades must be greater than the force of gravity.
2. When an automobile engine pushes a car up a hill, the force of the engine must be greater than the force of gravity.
3. The faster any object moves when it starts upward, the farther it will go before gravity stops it and then pulls it down.

[2] Ginn and Company, 1973.

Figure 4–2. Sketch in a third-grade textbook depicting the force of gravity. From *Introductory Level C*, Bendick, 1973, of the GINN SCIENCE PROGRAM, © Copyright, 1973 by Ginn and Company (Xerox Corporation). Used with permission.

Friction is then introduced as a "force which pushes something back." Children are encouraged to test the relative effects of friction with different surfaces around the house or outside. The reduction of friction through such means as lubricating surfaces and streamlining shapes are discussed. An experiment on air friction is suggested:

Drop a sheet of paper and watch it fall to the floor. What happens to it? Describe how it falls. The friction of the air slows it down. Crumple up a piece of paper. Drop it again. How does it fall this time? What can you do to the paper to make it fall even faster?

The interaction of various forces such as magnets, gravity, muscles, springs, and electricity are discussed and children are asked to identify various forces in their environment. The section on the interaction of force, motion, and friction is closed with a statement by a "scientist" pictured writing an equation on a chalkboard:

Today we believe that there are four basic forces in nature. You already know the first two—gravity is one. The second one is electromagnetic force. Magnetism and static electricity both come from force. Friction, spring forces, muscle forces, and all the other forces you feel (except gravity) come from electrical forces between atoms.

Two other forces have been found with giant atom smashers. These forces hold the tiny parts of atoms together. We know so little about them we just call them the strong force and the weak force. Let's hope we have better names for them by the time you grow up!

The reader is encouraged to use the checklist in Table 4–2 and the information in Part Two of this book to determine the general cognitive characteristics of eight- and nine-year-olds and compare these with the cognitive requirements of the three conceptual schemes just described as well as those that follow.

Concept 4: Molecules Change in Chemical but Not in Physical Changes.[3]

A mixture of sugar and iron filings is made and then separated by using a magnet. A filter is used to separate solid particles from the liquid in a solution. These and other examples of "physical changes" are used to show that molecules remain unchanged. Then some sugar is heated (in a picture) and it melts, boils, turns black, and so on, and the children are asked if the black substance is sugar. A ball and stick model (Figure 4–3) is then presented in picture form and the text and teacher explain that sugar molecules are made up of atoms: hydrogen, oxygen, and carbon.

The black substance that resulted upon the burning of sugar is identified as the carbon atoms (black in Figure 4–3), and oxygen and hydrogen atoms are red and yellow, respectively. It is further explained that water was given off during the heating process and water is actually made up of hydrogen and oxygen atoms. Other examples of physical and chemical changes are then described.

Concept 5: Air and Air Pressure

About 20 pages in the child's book are devoted to describing air as matter and air pressure as being capable of exerting forces. Since air (wind) can move things, it is described as matter. Various pictures and sketches depict air as matter, either pushing (wind) or resisting motion (parachute).

Air pressure is illustrated through pictures such as those in Figure 4–4. The teacher and the book explain that air pressure causes the can

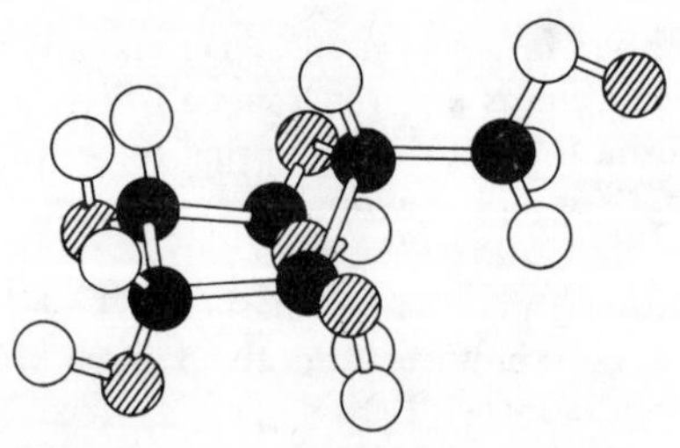

Figure 4–3. Ball-and-stick model of sugar in a third-grade textbook.

[3] Harcourt Brace Jovanovich, Inc. 1972.

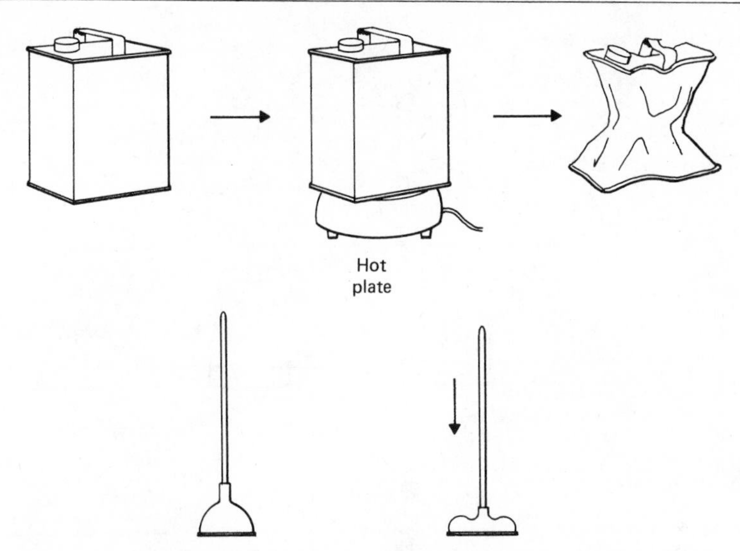

Figure 4–4. Illustrations of air pressure in a third-grade textbook.

to collapse and makes it very difficult to pull the plunger back up from the flat surface. Other examples of air pressure are presented to the third grader in pictorial form and some are shown in Figure 4–5. The teacher is provided with information to explain the phenomena in terms of air pressure.

The remaining few pages are devoted to explaining that air expands as it warms and contracts as it cools. One experiment designed to illustrate this principle for the children is shown in Figure 4–6. Two identical paper bags, *a* and *b,* are balanced as in illustration 1 and then bag *a* is filled with cold air from a container from the refrigerator. As shown in illustration 3, bag *a* then apparently weighs more because of the cold air "poured" into the bag. In illustration 3, the heat from the hot water rises up into the inverted paper bag and causes it to rise. This series of experiments is apparently intended to prove the validity of the "heaviness" of cold air relative to an equal volume of warm air.

The five concepts or conceptual schemes described as examples found in third-grade textbooks have not been chosen at random. Each one represents a "piece" of science that has been and continues to be extensively investigated by scientists, with resulting theories being developed and refined. One common element among the five "pieces" of science content is the *abstract* nature of all of them. Energy, sound waves, forces, molecular structure and changes, and air pressure are all abstract to the extent that our senses do not reveal their existence. All of the phenomena are really ideas or theories made up to explain the many

Figure 4–5. More examples of air pressure.

observations made by a large number of people, most of them called scientists. The nature of a wooden cube can be uncovered by seeing, touching, and manipulating so that sides, edges, and corners can all be related, weight can be sensed, texture and firmness can be explored, and the "cubeness" can gradually be internalized by the child. Even for this real object, however, it is no quick, easy process. Weight does not seem to be conceptualized, at least at a conserving level, until after the age of

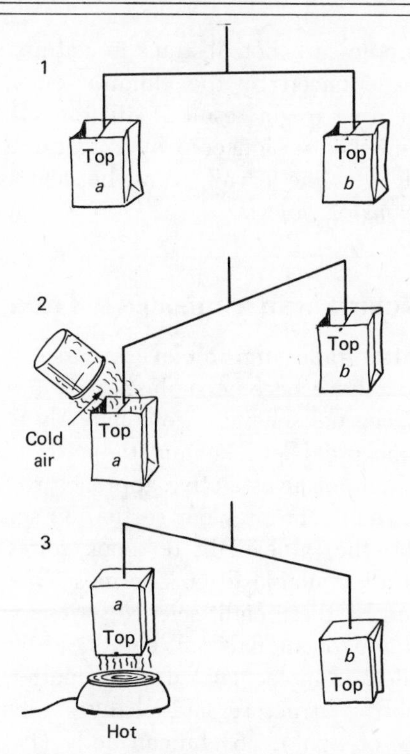

Figure 4–6. Experiment for third-grade children illustrating that cold air is heavier than warm air.

nine or ten years; volume is delayed slightly longer; metric proportions and quantitative relations among the length-area-volume properties are usually not available to children of elementary school age at all. In view of the limitations of children's thinking *with concrete objects,* it is a bit difficult to see how the eight- and nine-year-old would be capable of much more than rote learning when such abstract ideas as energy, wave motion, molecular structure and change, and air pressure are presented, and even then mostly in a verbal or pictorial (symbolic) manner.

Not all third-grade textbooks are filled with concepts as abstract and difficult to internalize as the preceding five examples. Most, however, reflect as the primary goal, information-giving. Quite often the information that is presented is in the field of concrete experience of the child and thus has a much greater chance of being understandable. Descriptions of animals and their young, characteristics of common plants, the growth of seeds and plants, characteristics of soil and rocks, characteristics of the human body, and numerous other examples of "science content" can be found in textbooks intended for eight- and nine-year-old children. The fact that these examples are not abstract in nature and that they have probably been experienced by the children does not change the fact that the "learning" that might result is still rote. If the cognitive development research and theory pioneered by Piaget and his colleagues are to be taken seriously, genuine learning must be seen as an active process of internalizing *actions on objects.*

Grade Five: Science Concepts and Children's Thinking

By the time children get to fifth grade (ten to eleven years), many of the science concepts in textbooks have become sophisticated and abstract. About a dozen units or conceptual schemes are commonly found in each science textbook. The conceptual level among the texts varies considerably, as does the emphasis on an investigative approach to finding out about the ideas presented. As in the previous section on science concepts in third grade textbooks, the same eight textbook series are used to "represent the market" of all available texts. The most common conceptual schemes, as represented by these eight science series for the elementary school are (1) plants and/or animals (all eight); (2) earth in space (six-of-eight); (3) atomic and molecular nature of matter and chemistry (five-of-eight); (4) earth's structure and changes (five-of-eight); (5) weather causes (three-of-eight); (6) human body (two-of-eight); (7) motion (one-of-eight). Other concepts are developed in the various series, but these are the most common, even though the nature of the presentation varies widely. Table 4–3 shows a brief listing of the various units, or conceptual schemes found in each of the eight texts.

The contents of each text are presented in a highly abbreviated form and really offer little opportunity for any valid analysis of the conceptual levels or the method of presentation. A careful analysis, however, does reveal wide differences in both conceptual level and method of presentation. Selected concepts from various texts are now described, in a form that approximates the manner in which they appear in respective fifth-grade books.

Concept 1: How Green Plants Get Their Energy

Since all eight textbook series had a chapter or part of a chapter devoted to plant growth, it seems appropriate to look at a "plant-related" concept first. Just preceding the section on "how green plants get their energy," this particular fifth-grade text showed a sketch of a magnified cross section of a leaf that showed various kinds of cells. It was noted that green matter called *chlorophyll* was contained within the inner cells of the leaf. Air flow within the leaf was also mentioned. Rather than try to paraphrase the specific section on "how green plants get their energy," the following information is presented just as it appeared in the student's text.[4]

Rich with chlorophyll, the long cells and spongy cells are the food factory of the leaf. Water enters these cells from the vein pipelines. Carbon dioxide, the "waste" gas that we breathe out and the gas that forms the bubbles in soda pop, flows into the leaf through the leaf openings. It then flows along the passageways and enters the long cells and the spongy cells.

In some way we do not yet fully understand, the energy of sunlight causes the carbon dioxide and water to form glucose, oxygen, and water. But, again, this happens only in cells that have chlorophyll.

This process of food making is called photosynthesis (fo′ tō sin′ the sis). Photo means "light" and synthesis means "putting together." So photosynthesis means "putting together in the presence of light." It is the way green plants get their energy.

In a word formula, we can show what happens during photosynthesis this way:

carbon dioxide (from air)	+ water (from roots)	+ chlorophyll	makes →	glucose (sugar)	+ oxygen (gas)	+ water vapor (gas)

Sugar made during photosynthesis is used by all living cells of the plant—in the leaves, stem, and roots. Sugar that is not needed by the plant right away is changed into starch and stored as food that the plant can use when photosynthesis stops, or when it slows down.

The book then instructs children to get three plants of the same kind and put one in sunlight, one in the same room but out of the sunlight, and the third in a dark closet. Records are to be kept concerning what happens.

The following page in the book describes how green plants use their energy and the process is termed *respiration.* Photosynthesis and respiration are compared in tabular form as in the following table.

More information on photosynthesis, respiration, and transpiration is presented on the pages that follow in the student's text, and the chapter is concluded by considering the world's food needs and how many calories of various kinds of foods can be produced per acre of land.

[4] From *Intermediate Level B,* by Gallant-Asimov of the GINN SCIENCE PROGRAM, © Copyright, 1973, by Ginn and Company (Xerox Corporation). Used with permission.

TABLE 4–3. **Summary of Science Units or Conceptual Schemes in Eight Fifth-Grade Texts**

Laidlaw	Holt	Rand	Silver Burdett	Leswing-Stone	Ginn	Harcourt Brace	Scott, Foresman
Living Things Characteristics of cells and cell activity. *The Earth* Formation, changes, and development of life. *The Universe* Characteristics of space and space travel. *Matter and Energy* Nature of electricity, light, and sounds. *Human Body* Eyes and light, ears and sound.	*Variety in Life* Characteristics of plants and animals; functions and reproduction. *Air, Water, and Weather* Sun's energy, air pressure, humidity, air masses, water pressure and currents. *Electricity and Magnetism* Electrons, fields, flow, potential differences, magnetic fields, electric current and magnetic fields, motors.	*Motion* Pendulums, falling bodies. *Heat Systems* Thermometers and temperature scales. *Weather System* Wind systems, cloud types, angle of sun's rays, air pressure, relative humidity. *Chemical Systems* Indicators, mystery powders, molecules and energy, chemical reactions, acids and bases. *Geology* Processes of change, wind water, earth-	*Building Blocks of Matter* Atoms and elements, chemical compounds, mixtures and solutions. *Earth* Minerals and rocks, metals and fuels, soils. *Plants* Plant parts, reproduction, algae, fungi, mosses. *Universe* Large distances, momentum, Milky Way, orbits, moon. *Energy* Kinetic and potential energy, force, work, forms of en-	*Spaceship Earth* Motion in space; time zones, solar system, seasons, climates. *Moon* Earth-moon motion, moon's characteristics, eclipses, tides. *Rocks and Minerals* Earth's interior, classes of rocks, formation, crystals. *Chemistry* Atoms and molecules, molecular structure, chemical changes, acids-bases, chemistry in nature.	*Leaves and Food* Types of leaves, photosynthesis, water. *Energy-Work* $W = F \times D$, levers, conservation of energy, kinetic and potential energy, heat energy. *Planets and Motion* Orbits, Kepler's laws, centrifugal and centripetal forces. *Human Body* Heart, blood flow, nervous system.	*Man and Earth* Force of gravity, mass and weight, actions and reactions, orbits, Newton's laws of motion. *Under the Surface* Weathering, fossils, earth structure, changes from within, rocks formed. *Chemical and Physical Changes* Molecules, chemical reactions, chemical structures, chemistry of the earth.	*Weather* Effects of heat, gravity, humidity, pressure. *Energy* Force and work, inertia, engines. *Plants* Reproduction and growth. *Earth and Moon* Rotation and time zones, centrifugal force, seasons, eclipses, tides. *Elements and Chemistry* Acid-base, oxidation, molecular structure, chemical compounds, chemical reactions.

Changes in Matter Physical and chemical changes, air as matter, atoms, molecules and heat.	quakes, heat faults. *Plant Processes* Growing plants, variables affecting growth, photosynthesis, how plants change.	ergy, conservation of energy, machines. *Sounds* Sound energy, speed of sound, frequency, pitch, acoustics. *Animals* Characteristics and classes. *Oceans* Bottom, currents, tides, as source of energy. *Cities* Pollution.	*Plants* Reproduction, mosses, ferns, algae, fungi, bacteria. *Animals* Similarities and differences, classes. *Plants and Animals Together* Populations; forests, deserts, etc.	*Particles in Motion* Matter as particles, temperature and speed of particles, convection. *Animal Development* Cellular development, insects. *Fossils and Creatures* Extinct creatures, fossils, adaption, evolution. *Webs of Life* Food chains, water and carbon cycles.	*Earth and Stars* Continuous motion, spectrum and light as waves or particles; spectral analysis. *Living Things* Cells and molecules, cell's energy, cell structure, reproduction, DNA. *Man and Environment* CO_2 cycle, molecules and eating, body cells and systems. *Evolution*	*Animals* One-celled, insects, higher forms. *Sound* As vibrations, as waves, frequency. *Burning* As chemical reactions.

TABLE 4–4. **Photosynthesis and Respiration As Compared in a Fifth-Grade Textbook***

Photosynthesis	Respiration
1. Goes on only in sunlight in plants whose cells have chlorophyll.	1. Goes on in the living cells of all plants and animals all the time.
2. Energy is stored in the form of food.	2. Energy in the form of food is used up.
3. Carbon dioxide is used by cells that have chlorophyll.	3. Carbon dioxide is given off by all living cells.
4. Oxygen is given off by cells that have chlorophyll.	4. Oxygen is used by all living cells.
5. Makes sugar.	5. Uses up sugar.
6. Adds mass or weight to the plant.	6. Uses up mass or weight.

* From *Intermediate Level B,* by Gallant-Asimov of the GINN SCIENCE PROGRAM, © Copyright, 1973, by Ginn and Company (Xerox Corporation). Used with permission.

To show the kind of variation in conceptual level among texts for fifth-grade children, the following concept is also concerned with plants, but the level of sophistication of the "content" varies widely.

Concept 2: The World of Plants

The word information, pictures, and suggestions for investigations are directed primarily toward identifying similarities and differences among plants. Attributes of size, color, shape, structure, means of reproducing new plants, and the like are compared and most of the efforts of the authors of this particular textbook, at least for this conceptual scheme, seem to have been directed at encouraging the children to become more aware of the variety of plants in their immediate environment and in some regions not necessarily available to the children, such as desert plants. Ferns, mosses, algae, fungi, and bacteria are described in the last few pages of the book and, particularly for the bacteria, children might have considerable difficulty in expanding their definition or class of plants to include invisible things that live on their own skin.

The obvious difference in the content of *concept 1* and *concept 2* is that in the latter almost all that is presented is directly observable by the child and abstract *explanations* or theories are avoided. Conversely, concept 1 is almost wholly directed toward explanations for phenomena not really "observable" in the sense used in concept 2. To complicate matters, the explanations presented were abstract and most certainly not within the conceptual framework of a ten- or eleven-year-old child. Photosynthesis, respiration, energy, chlorophyll, carbon dioxide, oxygen, cells, glucose, mass, and so on represent abstractions or "near-abstractions" that are not within a "testable" realm as far as a fifth-grader is concerned. Photosynthesis received more attention than the other concepts and was "explained" as a "reaction" in the chemical sense, except the components of the reaction were presented as words rather than CO_2, H_2O, and so on.

Following the description of a few more concepts in fifth-grade science texts, cognitive characteristics of fifth-graders are presented for the purpose of comparing the conceptual levels of the content with those of the children.

Concept 3: Atoms and Their Combinations

The opening chapter of this particular textbook deals with the "building blocks" of matter. The world of atoms and how they combine constitute the majority of the material. A "picture" of a crystal section of iron pyrites, magnified 44 million times, is used to show the position of the atoms in the crystal and children are then encouraged to think about "indirect" evidence for atoms by using "inference boxes." Air is defined as matter in the sense that millions of moving particles keep a balloon or a basketball or a tire blown up by striking the inside wall of the object.

Another investigation is then suggested to communicate to the children that small objects such as BBs have spaces between them in a container. This is meant to be indirect evidence that atoms of a gas or other matter have spaces between them.

Different types of atoms are then described as "elements" having different characteristics. Chemical symbols for a few dozen elements are presented and formulas for various compounds (for example, salt, $NaCl$; baking soda, $NaHCO_3$; sugar $C_{12}H_{22}O_{11}$) are described.

About twenty pages are devoted to atoms and their combinations, and a half dozen "investigations," meant to be used as a model for inferring the existence of atoms, are sprinkled throughout the pages. A few pictures and sketches also accompany the other written material.

Concept 4: Why Does the Weather Change?[5]

Just over 30 pages in this fifth-grade text are devoted to looking at causes for weather changes. Among the various concepts contained within the general conceptual scheme of "weather changes" are the following:

[5] Scott, Foresman and Company, 1968.

1. Temperature of air is determined partly by the temperature of land and sea surfaces.
2. Air touching a warmer object is heated by conduction.
3. Warm air is pushed up by cold air and convection currents are set up.
4. A surface that has absorbed radiant energy conducts heat.
5. Radiant energy may be absorbed or reflected when it strikes opaque material.
6. When radiant energy is absorbed by an object, it is changed to heat energy.
7. Air has weight because of the pull of gravity upon it.
8. Because gravity pulls air toward the center of the earth, air presses against the surface.
9. Pressure of air is not the same everywhere on the Earth.
10. Cold air weighs more than warm air.
11. Cold air has high pressure and warm air has low pressure.
12. Air exerts an equal amount of downward, upward, and sideways pressure at a given point.
13. Winds are caused by differences in air pressure.
14. Differences in air pressure are caused by unequal heating of air.
15. The greater the temperature difference between two air masses, the greater is the difference in pressure.
16. When air has all of the water vapor it can hold at a certain temperature, it is saturated.
17. When unsaturated moist air is cooled enough, it becomes saturated.
18. Fog is a cloud formed when saturated air near the ground is cooled.
19. When clouds are cooled, tiny particles of moisture join together to form drops large enough to be attracted by gravity.
20. Snow forms when saturated air is cooled below the freezing point of water and water vapor changes to ice crystals instead of condensing to form a liquid.
21. If a cold air mass is moving in and pushing a warmer air mass out of the way, there is a cold front where they meet.

The preceding concepts are presented in the student's text mostly in the form of words, with accompanying sketches and pictures. Because of the rather abstract nature of most of the concepts, the sketches usually depict *models* intended to represent the various causal factors involved in weather changes. One example of a sketch that depicts a representational model is shown in Figure 4–7.

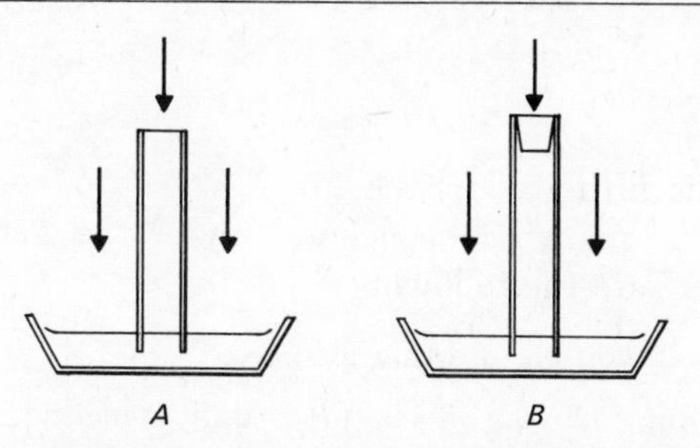

Figure 4–7. Sketch in a fifth-grade science text depicting the downward pressure of air due to gravity. (From *Science Is Discovering* by Glenn O. Blough, J. Stanley Marshall, James B. Bailey, and Wilbur L. Beauchamp. Copyright © 1965, 1968 by Scott, Foresman and Company. Reprinted by permission of the publisher.)

The information accompanying the sketch explains what to do and how gravity, air pressure, and water pressure interact to keep the water up in the glass tube in *B* (closed) but not in *A*. This demonstration is supposed to represent the general effects of gravity on air masses, and the child is expected to generalize the demonstration represented by Figure 4–7 to causal effects in weather changes.

While the ten- or eleven-year-old fifth-grade child is cognitively far advanced from the six- or seven-year-old first-grade level, many limitations in his or her thinking are apparent. Formal or abstract thinking is still not available, although a few children are exhibiting signs of the beginnings of such operations. Classification is now firmly established, and many children are capable of fairly sophisticated schemes. Stable class inclusion ideas and the ability to construct complex hierarchical classes are, however, still beyond most ten- and eleven-year-old children who are more likely to search for physical causality explanations and reject animism and artificialism in causal considerations. "Pieces" of a system are still not related as a systematic whole and well-mixed variables cannot be separated. Experimental variability is seen typically as a result of multiple causes, and no systematic efforts are made to isolate and control variables. Such formal or logical schemas as chance, correlations, proportions, and combinations are not operable, and for most children, still years away. Thinking is still a derivative of the child's own actions on real objects that represent themselves rather than some larger, more abstract system.

The reader is encouraged to attempt to determine the appropriateness of the various concepts presented in this section in terms of the general conceptual level of fifth graders.

Concept 5: Energy and Work[6]

Almost 50 pages in this fifth-grade book are devoted to energy, forces, and work. Among the many "subconcepts" under the general heading of "energy and work" are the following:

1. When you push or pull, you are exerting a force.
2. Once you start something moving, it continues until something gets in its way.
3. Work equals force times distance: $W = F \times d$.
4. Work is done only if an object moves.
5. A lever transfers a force from one place to another and changes its direction.
6. A lever with equal lengths on either side of its fulcrum changes the direction of a force.
7. When a substance or force can only be transferred, when more of it cannot be created, and none of it can be destroyed, it is said to be conserved.
8. The kind of energy that is connected with motion is called kinetic energy.
9. The higher a ball is raised, the more work is being done on it, and the more potential energy it has.
10. The higher you lift a ball then, the more kinetic energy it gains when it drops, and the more potential energy it must have had before it dropped.
11. When a moving object disturbs atoms in the air, each atom gets some share of additional kinetic energy and moves a little faster.
12. Scientists have now discovered that some kinds of energy can become matter. Some very high energy radiation can turn into a pair of tiny particles that are even tinier than an atom.

Most of the previous 12 statements are in the same form as they appeared in the fifth-grade text. Obviously, these statements are taken out of the context in which they appeared and this tends to add to the apparent degree of difficulty and abstractness. The context, however, was primarily other words explaining the 12 concepts and various illustrations and pictures meant to clarify or simplify the concepts. Because of the abstract nature of such ideas as energy, force, and work, investigations or demonstrations suggested in the text are necessarily "inferential models" meant to reduce the abstractness of the ideas.

[6] Ginn and Company, 1973.

It is left for the reader to estimate the conceptual level of the "energy and work" unit and compare it with general cognitive characteristics of ten- and eleven-year-old children.

Concept 6: Movements of Earth and Moon

As with the previous conceptual scheme of energy and work, this rather broad concept of movements of the Earth and the moon is subdivided into many smaller concepts:

1. The Earth rotates on its axis, an imaginary line running through its center, once every 24 hours.
2. Meridians are imaginary lines on the Earth passing through both poles.
3. The sun is highest in the sky at different times of the day in different places because the Earth rotates.
4. By sun time, it is noon on any meridian when the meridian is carried by the Earth's rotation under the sun's most direct rays.
5. The Earth's orbit is about 93 million miles from the sun.
6. The sun's gravitational pull keeps the Earth in its orbit so that it does not fly off into space.
7. Centrifugal force keeps the Earth from being pulled into the sun by the force of the sun's gravity.
8. Seasons are caused by the revolution of the Earth and the inclination of its axis. The amount of sunlight that certain parts of the Earth receive varies from season to season.
9. The moon is kept in its orbit by the pull of the Earth's gravity balanced by centrifugal force.
10. Orbits of the Earth and moon usually do not lie in the same plane.
11. The moon's gravitational force acting on the Earth is the chief cause of tides.

The only type of investigations or demonstrations that are possible with most of these concepts are those that deal with spatial relationships. Projection of shadows, relative position and motion and angles and planes can be illustrated with globes and the like, but gravitational forces and centrifugal forces cannot be so easily demonstrated. Children must be capable of translating "static" pictures or sketches into "dynamic" systems that involve rather complex relative positions and relative motions (rotations and revolutions). Needless to say, very few ten- or eleven-year-old children are capable of such mental gymnastics.

[7] Scott, Foresman, 1968.

It might be appropriate at this point to review some general cognitive characteristics of fifth graders to allow the reader to compare them with the apparent conceptual levels of the various concepts that have appeared in this section of the book.

General Cognitive Characteristics of Children in Grade 5

By the age of ten or eleven years, most children have advanced quite far in classification and seriation abilities. Class inclusion, multiplicative classes, null class, and hierarchical classification are not highly functional, however, so complex types of classification are beyond most children in fifth grade. Their ability to represent (through drawings) various two- and three-dimensional objects is fairly good, except for more complex objects. Related to this, the children are much more likely to be able to conceptualize more than one perspective, but again only for simple objects. The quantitative abilities associated with Euclidean geometry are generally not available. Length and distance are measured with some accuracy but area and volume measurements are inaccurate and primitive because of a lack of ability to conserve the quantities, especially volume. Their ability to conceptualize and use a coordinate system is not generalized and, thus, quantitative graphic representations involving vertical-horizontal coordination, angle measurement, and the like are still at a level beyond most ten- and eleven-year-old children. Time and motion studies are beginning to make more sense, but only for simple, straight-line (linear) motion. Comparing relative speeds, dealing with acceleration, and the like are as yet too difficult. Many children in fifth grade are capable of simple weight conservation, but density, weight-distance equilibration, and other "two-quantity" concepts are beyond the developmental levels of these ten- and eleven-year-old children.

Schemas associated with more formal, abstract thought such as probability, correlations, proportions, combinations, separating and controlling variables, and propositional logic are considerably above the cognitive levels of the great majority of ten- and eleven-year-old children. Experimental variability is seen as a result of many causes and, thus, no genuine efforts are made to search for causal effects. Conservation is still limited to those areas that are easily tested, and thinking remains a derivative of the child's own actions on objects. "Models" that are meant to represent abstract systems (for example, atoms, molecules, energy change, weather causes) in fact only represent the concrete reality of the objects themselves. In general, causal explanations that are commonly associated with phenomena in science cannot be conceptualized by the ten- or eleven-year-old child because of the requirement that such conceptualizations must be made up in the mind of the child without any corresponding concrete reality.

The checklist format used to summarize conceptual levels of first- and third-grade children is again shown in the following table, for fifth-grade children. The reader is again reminded of the 2- to 3-year range of developmental levels that can be expected at any grade level.

Before going on, the reader is encouraged to locate some elementary science texts (preferably teacher's editions) and try to determine the general conceptual levels. The concepts described in Part Four of this book, thus far, are not necessarily representative of all science concepts in elementary textbooks. These concepts were chosen primarily to make

TABLE 4–5. **A Checklist for Use in Analyzing Science Concepts Available to *Fifth Grade* (Ten and Eleven Years) Children**

Description or Requirement	Likelihood of Availability Yes	Some	No
1. Actions on objects approach required.	x		
2. Objects represent only themselves (no models).	x		
3. Multiple classes and stable classification criteria.	x		
4. Hierarchical classes.		x	
5. Seriation by length or width.	x		
6. More complex whole number operations.	x		
7. Decisions governed by perceptions.		x	
8. Accurate measurement of length.	x		
9. Estimate distances.		x	
10. Causal thinking.	x		
11. Coordinated perspective.		x	
12. Objects can be interrelated as a whole.	x		
13. Functional conception of area.		x	
14. Functional conception of volume.		x	
15. Generalized reference system.		x	
16. Functional conception of time and motion.		x	
17. Weight conservation.		x	
18. Recognize contradictions.		x	
19. Imagine "unseeable" objects (atoms).			x
20. Imagine events not experienced.		x	
21. Separate variables.			x
22. Control variables.			x
23. Identify all combinations.			x
24. Probability.			x
25. Ratios.		x	
26. Proportions.			x
27. Deductive logic in proofs.			x

the point that much of what you might find in state-adopted or locally adopted science textbooks for elementary children is not necessarily conceptually appropriate for the children. The nature of the concept *and* the method(s) used to facilitate learning must be considered before making decisions about the adoption and use of curriculum materials.

Some of the curriculum projects centered around conceptual schemes such as Science Curriculum Improvement Study (SCIS), and Conceptually Oriented Programs in Elementary Science (COPES) can also be analyzed in terms of conceptual level. Teacher's guides describe the concepts to be learned, methodology, equipment, and the like, and a careful analysis of the materials can yield valuable information concerning the appropriateness in terms of cognitive levels of children. Before going on to briefly discuss examples of postelementary science texts, portions of some elementary texts are analyzed in terms of the ratio of information presented to the suggested investigations.

Quantitative Analysis of Science Texts

One fairly easy, straightforward method of analyzing science texts and other materials is to determine the ratio of information presented to investigations suggested for the children. All *demonstrations* by the teacher might be considered a part of the information presented rather than as a part of the investigations for the children. However, for clarification purposes, two categories of "information presented" could be used, since a demonstration is a type of information presentation. Also, under student investigations, two categories might include those suggested for classroom or schooltime activities and those suggested as home experiments.

The following two tables are intended to show examples of instruments that can be used to analyze science texts according to two categories: (1) information, and (2) demonstrations. Each category is divided into at least two subcategories, concrete and abstract. In the first category, information, Table 4–6 shows five means of presenting information in books as words, pictures and sketches, diagrams, tables, and graphs. These five means of presenting information account for nearly all of that which is usually found in science texts for children.

Concrete refers to those things that are a part of the child's environment and that can be physically sensed by the child (for example seeing, touching, smelling). *Abstract* refers to those things or ideas that cannot themselves be directly sensed. Light can be sensed but light waves cannot (remember, we are talking about elementary schoolchildren here). The greenness of a leaf can be sensed but the "chlorophyll" cannot. The liquid inside a leaf can be sensed but photosynthesis

TABLE 4–6. A Means for Analyzing *Information* in Science Texts for Children

	Concrete	Abstract
	Per Cent	*Per Cent*
Words Pictures and		
Sketches		
Diagrams		
Tables		
Graphs		

cannot. Perfume can be sensed but "molecules" cannot. A fossil can be sensed but evolution cannot, and so on. Similar comparisons can be used for pictures and sketches, diagrams, tables, and graphs. Anything that does not allow for verification (by the child) in a concrete sense must be considered as a type of abstraction.

With these thoughts in mind, the reader is encouraged to try some analyses of science information contained in materials for children. Instead of using only a two-category system, concrete or abstract, it might be better to extend the system to three or four or five categories. Perhaps "concrete" could be labeled 1, abstract could be 5, and 2, 3, and 4 could represent "degrees" of abstractness. This complicates the analysis and interpretation of results, but nevertheless is probably a more valid way of analyzing content information, since there *are* many different levels of difficulty.

The amount of material in most texts that is devoted to teacher demonstrations is much less than for information by words, pictures, diagrams, and the like. Also, "student investigations" often become "teacher demonstrations" because of the difficulty of getting anything but the simplest and least expensive science equipment into an elementary school classroom.

Science curricula for children vary widely in the quantity and quality of suggested student investigations. It is this aspect of the "contents" of a text that would seem to follow the implications of Piaget's

TABLE 4–7. A Means for Analyzing Demonstrations in Science Texts for Children

	Concrete	Abstract
	Per Cent	Per Cent
Demonstrations		

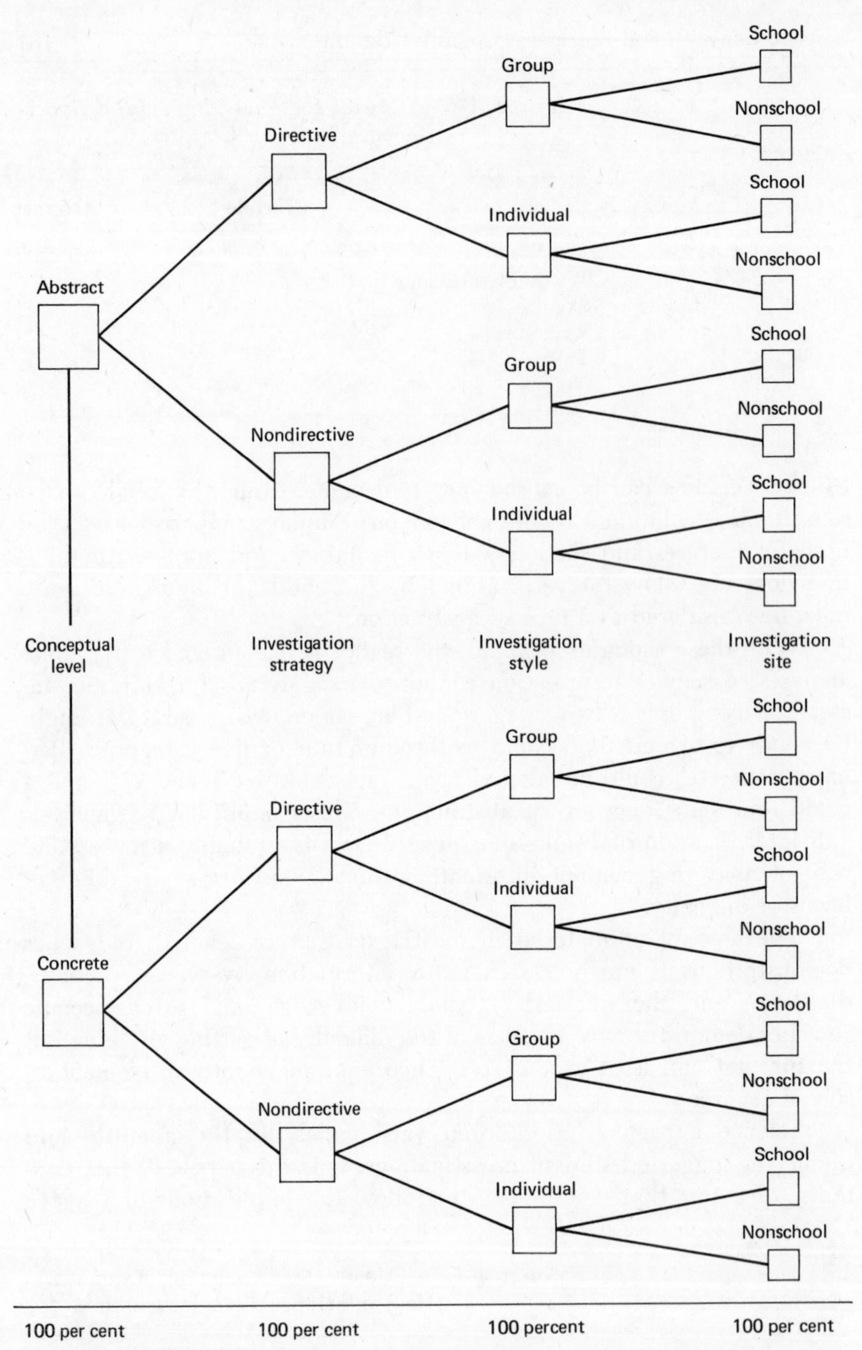

Figure 4–8. A means for analyzing the nature of *investigations* in science texts.

"actions on objects" theory of cognitive development. Internal or mental operations are derived from the external actions of the child. Investigations can provide a means for an "actions on objects" approach for the elementary schoolchild, but the nature of so-called "investigations" varies widely among different curriculum materials.

Figure 4–8 shows a means for determining the nature of investigations for students. Although the number of criteria that could be used to analyze the nature of "investigations" intended for children is large, four in particular seem important and they are included in Figure 4–8. The four criteria are *conceptual level* (concrete to abstract), *instructional strategy* (directive to nondirective), *investigative style* (group to individual), and *investigation site* (school or nonschool).

Especially for "conceptual level" and "instructional strategy" it is an oversimplification to divide each into only two categories. There is a continuum of "concreteness" to "abstractness" just as there is in "directive" to "nondirective" teaching strategies. For the purposes of a kind of "rough analysis," however, these two categories can be divided into a forced choice position, as long as the reader is aware of the shortcomings of such an approach. The "model" in Figure 4–8 is intended primarily as an example of one way an investigation in a science text or other science curriculum materials could be viewed.

Science Concepts in Postelementary Texts

Part Four has thus far considered science concepts found in elementary texts and how the concepts compare to the general conceptual level of the children. It is quite clear that children of elementary school age are not capable of abstract or formal operational thought as described in Part Two. After the age of eleven or twelve years, however, Piaget has shown that some children begin to reflect an ability to engage in formal, abstract thought. In fact, upon briefly reading his theory of cognitive development, it is fairly easy to assume that most or all children attain the ability to engage in abstract thought by the age of fourteen or fifteen years. An increasing number of studies are suggesting, however, that most adolescents and adults do not, and perhaps cannot, hypothesize about possibilities, separate and control variables, and use proportions or correlations or probability in problem-solving situations. If these studies accurately reflect the "typical" cognitive functioning of postelementary schoolchildren, then a large part of what appears in many, in fact most, science textbooks is at a conceptual level that is too high. Conceptual schemes such as atomic-molecular nature of matter, matter-energy conservation, genetic nature of life, atomic and nuclear interaction, and others are by their very nature theoretical and abstract rather than observable and confirmable by direct experimental evidence gathered by the student.

One would expect that very few children in the age group of eleven to fifteen years would be capable of logical, abstract thought since, according to developmental theory, it is during this time that the potential for such thought develops. Even after the age of fourteen or fifteen years, there seems to be only a minority of persons who are capable of such mental operations as proportions, probability, all possible combinations, and eliminating contradictions (refer to the section on formal thinking in Part Two). The discipline called *chemistry* usually involves content that is highly abstract (atoms, molecules, orbitals) and highly quantitative (balancing equations, calculations of energy exchange, comparing atomic weights, densities). *Models* representing hypothetical objects and events typically are introduced very early in the curriculum materials and the students are directed to engage in activities that are abstract and quantitative.

The quantity of science concepts that appear in postelementary science texts is extremely great and for that reason no systematic analysis of representative curriculum materials is presented here as was done earlier for elementary science texts. The reader is encouraged to locate some fairly recent postelementary science tests and attempt to make some determination of the general conceptual levels, percentage of pages devoted to investigations, type of investigations (for example, concrete to abstract, individual to group) and any other analyses that would help to characterize the materials.

science processes and children's thinking

Introduction

The previous part of this chapter emphasized science as "content" and used textbook material as example content (concepts). Science can also be interpreted as "process" and it is this view that is described in some detail here. The curriculum material used to exemplify a process interpretation of science is Science: A Process Approach (SAPA). This elementary science curriculum project of the 1960s is used because it so clearly defines science as a set of processes, while de-emphasizing the content position. It is because this distinction between content (concepts) and process is so clearly made by the developers of SAPA that this curriculum is analyzed in more detail than other possibilities, such as Elementary Science Study (ESS) and Science Curriculum Improvement Study (SCIS). The reader should not interpret this emphasis on SAPA as meaning that it is somehow "better" than other possibilities. "Better" has meaning only to the extent that criteria have been clearly established and an objective analysis of some data allows for such a comparative judgment.

The intent here is to differentiate clearly between science as content (concepts) and science as process. It is hoped that this clarification will place the following description of science as a process in the proper perspective.

Most curriculum materials in science that are available today are centered around concepts. One curriculum development project in particular, Science: A Process Approach[8] focused its attention around science as a set of processes. The developers of this elementary science curriculum took a *drastic* step that was practically unheard of in curriculum development circles. They eliminated reading materials! Subject matter content (concepts) was not to be substituted for science. Instead, *processes* were used as a definition for science and the "curriculum" consisted of doing things the developers believed represented the real essence of science. "What scientists do is what defines science" is perhaps another way of looking at the intentions of the developers of SAPA.

Eight processes were emphasized for grades K–3:

1. Observing.
2. Classifying.
3. Using numbers.
4. Measuring.
5. Using space–time relationships.
6. Communicating.
7. Predicting.
8. Inferring.

For grades 4–6 the following "integrated" processes were emphasized through various student and teacher activities:

9. Defining operationally.
10. Interpreting data.
11. Formulating hypotheses.

[8] Published by Xerox.

12. Controlling variables.

13. Experimenting.

The SAPA group believed that these 13 processes defined the domain of science for elementary schoolchildren. The reader might have noticed that some of the science processes are very similar to the developmental processes discussed in Part Two (classifying, space/time relations, controlling variables). Defining science as a set of processes by no means eliminates the learning of concepts. Processes do not occur in a vacuum. The eight "basic" processes that appear in the K–3 part of the curriculum use, primarily, physical systems. Some examples of how these "physical systems" can be used to emphasize "science as process" are now described in some detail. Any considerations of the teacher's role will be postponed until Part Five, but attempts are made to compare the conceptual or developmental level of the curriculum materials with the "expected" range of cognitive abilities for a specified grade level.

Science Processes at the Kindergarten Level (SAPA)

The processes of observing, classifying, using space/time relationships, using numbers, and measuring are introduced at the kindergarten level. Twenty-two lessons are built around the processes, using mostly static sets of objects varying according to color, shape, size, temperature, sound, length, and so on. Observing and using space/time relationships constitute about two thirds of the total number of process lessons. A few examples of lessons that emphasize a given "process of science" are described with some brief comments about the general cognitive characteristics of five- and six-year-old children.

Observing

Eight lessons on observing are contained in the teacher's guide beginning with identifying and naming primary and secondary colors and ending with identifying and naming the taste of objects as sweet, sour, or salty. In the third lesson on observing, children work with thermometers in measuring the "temperature" of liquids that have been precooled or heated to varying degrees. They also use the thermometer in "observing" the temperature during the day and from day to day. It is expected that each child (or a large majority of them) will be able to learn to use a thermometer and read the scale to determine the temperature.

In another observing lesson, the children distinguish between sounds in terms of volume, duration, and pitch. The teacher asks children about loudness and softness of various sounds made by either children or the teacher. Sounds from paper, pencils, marbles, musical instruments, drinking glasses, and any other "sound-makers" are "observed" and discussed. It is expected that each child will be able to describe different sounds in terms of loudness, duration, and pitch.

A fifth lesson on observing involves identifying and naming color changes. Children are shown that mixing blue and yellow food coloring together results in a green liquid. They are then given a blue and a yellow crayon and a piece of white paper and asked to try to make a green color on the paper. The children are then shown that a piece of cloth can change colors if it is dyed with an indicator and then dipped into a solution such as sodium bicarbonate. Each child is then supplied with a piece of Congo-red paper (indicator) and the children are told to dip a paste stick in vinegar and touch it to the paper. The resulting reaction causes a color change from red to blue. It is expected that each child will be able to match colors and correctly state whether or not a color change occurs.

The previous three examples are all "observing" lessons and, according to the SAPA developers, should help the children develop the process skill of observing. It is also fairly obvious that other conceptual processes easily could be involved in addition to observing. Classifying by size, shape, texture, and the like as well as counting and using space relationships very likely would be involved. Piaget has emphasized the interrelatedness of the mental operations and schemas in the developing organism, and this "systems approach" principle certainly holds when interacting with sets of science-type materials. Just because a lesson is identified as an observing lesson does not mean that children's mental operations are limited to that process.

Using Space/Time Relationships

A second kind of process skill that can be associated with "doing science" is that of using space/time relations. It was emphasized earlier in Part Two that the development of topological, projective, and Euclidean concepts in geometry are a crucial component in the general development of logical thinking. Six lessons on space/time relations are described in the teacher's guide to SAPA, Part A (kindergarten) and a few examples are presented here for purposes of illustrating this process skill of science.

The first lesson involves identifying and naming such two-dimensional shapes as triangles, circles, squares, rectangles, and ellipses. Children are given at least one shape and are told its name. They then practice drawing around the shape. Each shape is compared to various objects in the room and then common objects such as a wagon are identified by the teacher to help children see combinations of "basic" shapes. An optional activity is suggested for the teacher to use with more "mature" children. Figure 4–9 shows a simple device for making ellipses. The teacher is asked to decide (on some basis) which children might be ready to construct ellipses using the tacks or pins and string technique.

From all of these activities each child is expected to learn the names of the common shapes and to identify the common two-dimensional shapes in his/her environment.

Another process lesson on space/time relations involves identifying and demonstrating the directions of up, down, forward, back, left, right. The teacher discusses the words *up* and *down*, *back* and *forward*, and then directs the children in corresponding movements in a "Simon says" game. Through similar means, right and left are introduced and a generalizing experience of going on a treasure hunt is used. From these experiences, the children are expected to learn the various directional concepts.

A third example of a process lesson (SAPA) on space/time relations at the kindergarten level involves constructing spatial arrangements of objects from pictures and diagrams. The teacher draws various shapes on the chalkboard and marks an *X*, which indicates the position a child should take in a corresponding shape on the floor, as in Figure 4–10. Various shapes are used, and different children get a chance to locate themselves in the position indicated by the *X*.

Children are then asked to use themselves to form various shapes in the classroom. As a generalizing experience the teacher draws the classroom plan on the chalkboard, points to a given position, and asks the child in that position (chair at a table) to go to a new location, which the teacher identifies again by pointing to another location in the

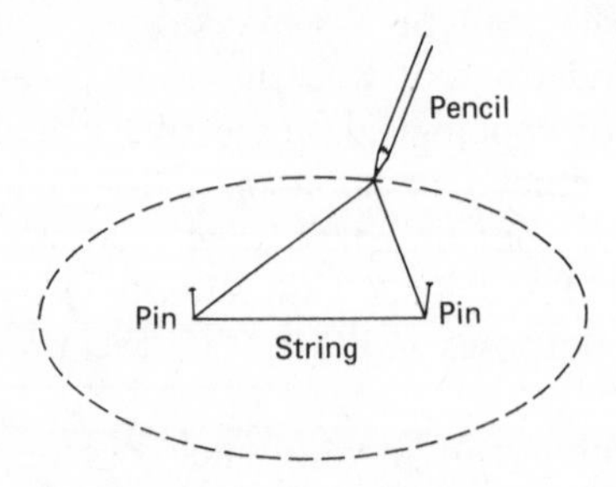

Figure 4–9. Constructing an ellipse from string, pins, and cardboard.

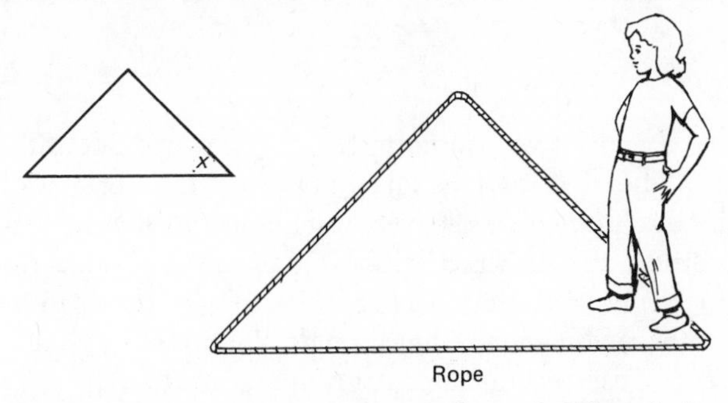

Figure 4–10. Locating positions from chalkboard sketches.

sketch. From these experiences, it is expected that each child will be able to arrange objects in the shape of a triangle, square, rectangle, and the like and to draw pictures of shapes of various arrangements of objects.

Using Numbers

A third process skill that is a part of the kindergarten program is using numbers. The first of four lessons throughout the program on using numbers involves identifying sets and their members. The teacher shows the class sets of shapes and other objects and emphasizes the terms *set* and *member*. Children are asked to identify sets of objects in the classroom and eventually the teacher tries to introduce the concepts of a single member set and a no member set (empty set). From these and related experiences, the children are expected to identify various types of sets, including a set with only one member and an empty set.

In a later lesson on using numbers, the teacher is directed to provide experiences for the children that involve identifying and naming the ordinal relationships among the numbers 1 to 5. Large note cards, each with a numeral from 1 to 5 printed on it, are used by five children in the class to "make a train." The teacher also asks some children to order their families according to firstborn, secondborn, and so on. Other activities suggested by the teacher include looking at page numbers in books and numerals on a calendar, and ordering, numerically, pictures of animals. As a generalizing experience, it is suggested that the teacher tell a story such as "Jack and Jill" and then ask some children to order a few pictures that represent certain aspects of the story. As a result of such experiences, it is expected that the children will be able to identify and name the position of an object or event in a sequence of five, using the words *first, second, third, fourth,* or *fifth.*

Classifying

Only three of twenty-two kindergarten lessons are identified as classifying lessons. The first two lessons introduce the term and ask children to classify such objects as leaves, nuts, and the like in first one way and then another. In the third lesson on classifying, children are asked to bring in a pet and the teacher tries to get them to notice similarities and differences among the animals. Pictures are then used by the teacher to suggest certain logical groups according to sex, size, and other features of the animal shown in the pictures. As a result of such experiences, it is expected that the children will be able to state how some animals are similar and/or different and then construct a classification of pictures of animals based on physical or behavioral attributes.

Measuring

In the single lesson on measuring, sets of wooden dowels of various lengths are used by the teacher to get the children to compare lengths and order the eight sets by length. Some of the dowels are then used to measure some preselected objects (paper strips) that correspond to the lengths of many of the dowels. A generalizing experience is suggested where the teacher tells groups of children to get objects in the classroom and then to order them from longest to shortest. Based on these experiences, it is expected that the children will be able to order objects by length and to measure by comparing object A with object B in terms of the length of a third object C, all in terms of length (for example, if $A = B$ and $B = C$ then $A = C$).

The five science processes at the kindergarten level (SAPA) have been identified as observing, using space/time relations, using numbers, classifying, and measuring. Science concepts such as those identified earlier from textbooks are not a part of SAPA even though, as we see, science content (concepts) are necessarily present. Instead of using a collection of leaves for the purpose of naming and learning other facts, a process approach uses the leaves to emphasize grouping or counting or measuring, and so on. Apart from names, the children cannot help but learn concepts about leaves such as variations in shape, vein structure, colorings, stem size, texture, association with certain trees (if collected by the children) and a host of other "leaf" concepts. Given opportunities to interact with objects in the environment, the child naturally tends to probe and learn about the objects and various relationships that grow out of an "actions on objects" approach. The extent to which children have opportunities to internalize their actions is, according to Piaget's theory of adaptation and cognitive growth, directly related to the development of conceptual schemas that allow for the understanding of what are called science concepts.

Science Processes at the First-Grade Level (SAPA)

A sixth process, *communicating*, is introduced at the second level of SAPA. In all, 26 lessons on process skills are contained in the teacher's guide in the following quantities: measuring (6), observing (5), using space/time relationships (5), communicating (5), using numbers (3), and classifying (2). As in the previous section on science processes for kindergarten, a few examples of each of the six process skills are presented to assist the reader in developing a "feeling" for the emphasis of science as *processes* rather than as *concepts*.

Observing

The first of the five lessons on observing involves the use of the senses to identify objects or changes in objects. The second lesson, which is described here, is somewhat different than most in that the lesson can continue over a period of many weeks if the teacher and the children decide to do so. Weather charts are used to record data on temperature using a thermometer, as well as clouds, wind, and other weather conditions. It is suggested in the teacher's guide that the teacher review with the children the function and operation of a thermometer in preparation for record keeping using the weather chart. Figure 4–11 shows the symbols the children are supposed to use in recording the specific weather conditions, and Figure 4–12 shows the weather chart that is provided for them to keep daily records.

As the charts are filled in by the children, the teacher is expected to discuss aspects of the data in an effort to help the children compare weather conditions from day to day or week to week. From these experiences the children are expected to be able to (1) construct weather conditions on a chart using standard symbols, (2) distinguish between weather conditions by reading a chart, (3) compare weather conditions from day to day in terms of "warmer than" or "cooler than" or "cloudier than," and so on, and (4) state the temperature in degrees using a thermometer.

This observing lesson consists of a great deal more than simply

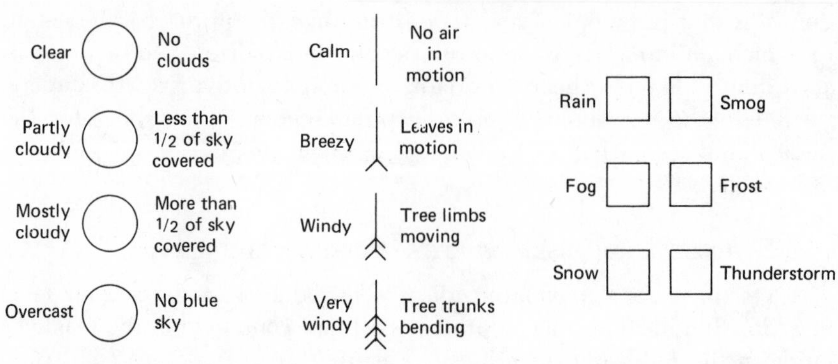

Figure 4–11. Weather symbols suggested for use in the first grade (SAPA).

observing. Extensive use of symbols is apparent throughout the experiences in the use of a temperature scale on thermometers, standard weather symbols, and the chart for recording the symbolic weather information. The extent to which the six-year-old child can make sense out of much of the observing lesson is certainly open to question. Reading a chart such as the one in Figure 4–12 requires that columns and rows be coordinated so that a given box can be used to draw meaning about weather conditions on a particular day. Also, days of the week must be related to one another in a sequential manner and associated with a "date" of the year. Information in Part Two on prelogical thought suggests that coordination of the vertical and horizontal (columns and rows) is not developed until well after the age of six or seven years. Coordinating the data on various weather conditions into a systematic whole on a chart appears to be incongruous with many of the results reported in Part Two for six- and seven-year-old children.

Other process lessons in SAPA for grade 1 on observing include properties of magnets, color changes in plants, and mold gardens.

Day Date	M.	Tu.	W.	Th.	F.	M.	Tu.	W.	Th.	F.	M.	Tu.	W.	Th.	F.
Temp.															
Clouds	○	○	○	○	○	○	○	○	○	○	○	○	○	○	○
Wind	│	│	│	│	│	│	│	│	│	│	│	│	│	│	│
Other	□	□	□	□	□	□	□	□	□	□	□	□	□	□	□
Name															

Figure 4–12. Weather chart for use in the first grade (SAPA).

Using Space/Time Relationships

Five of the twenty-six process lessons in grade 1 deal with using space/time relations. The content or concepts used to involve the children in this process include (1) symmetry or geometric shapes, letters, fruit, people, and other common objects; (2) general shapes (oval, circle, rectangle, triangle, and others) of animals; (3) casting shadows of three-dimensional objects; (4) comparing angles; and (5) using time-measuring devices to distinguish between short time intervals. Although the content varies widely, the common process of space/time relations ties all of them together. Science as a process or set of processes uses content merely as a vehicle for furthering the process in question. The learning goal involves advancement of an ability to *do* something rather than to *say* something (except where communication as a process is a goal in itself).

The last two process lessons on space/time relations identified, comparing angles and using time-measuring devices, are described as examples of this basic process for first-grade children.

Comparing angles involves identifying, naming, and comparing relative sizes of angles by placing one angle on top of another. The teacher begins by drawing a familiar street intersection on the chalkboard and leading a discussion on directions and angles. This is later expanded to include the classroom and the school. Wire outlines of a square and a rectangle, paper shapes of the same plus triangles, pipe cleaners, coordinate paper, and other materials are then used, all for the primary purpose of comparing angles. The coordinate paper is used by the teacher to direct the children to trace a path according to certain instructions (that is, move ahead three steps, move right five steps, move right again and go eight steps). Other activities that are suggested include (1) search through old magazines for pictures containing angles, (2) check the angles made by doors, and (3) make "body angles" with arms and legs. As a result of the various experiences with angles, it is expected that the children will be able to exhibit certain learned behaviors.

Another space/time process deals with the use of various devices that can be used as "timers." The teacher asks leading questions about duration of events and then uses a record player and an old record that has a scratch so that an audible "tick" occurs for each rotation. A sand timer is then made by using a funnel and a glass bottle. Two children help the teacher while the others watch, and then eight groups of about four children each use similar timers to compare various events. Water timers, a pendulum, and a burning candle are all suggested as examples that might also be used by the children to further explore the duration of familiar events.

Using Numbers

The three lessons that emphasize this process are geared toward helping the children learn how to (1) name ordinal positions on the number line for integers from 10 to −10; (2) name the whole numbers from 0 to 99; (3) add any two whole numbers whose sum is less than 100.

Classifying

Both of the classifying lessons for first grade involve the construction of a single-stage classification system. That is, a group or class is determined by only one attribute such as shape, color, texture, odor, or size. Pictures as well as various objects are used by the teacher to see that the children attain the skills outlined in the teacher's guide.

Measuring

The measuring process is emphasized most for first graders (six lessons) and includes skill lessons such as:

1. Finding the length of objects using a "unit" measure.
2. Comparing volumes of containers using a "unit" measure.
3. Measuring length in centimeters, decimeters, meters.
4. Comparing the weights of various objects using equal-arm balances.
5. Measuring area using a "unit" measure.
6. Measuring the growth of sprouting seeds and recording the data in table form.

The conceptual levels of *processes* can vary just as greatly as with *concepts* and it is difficult to ascertain such levels from summary descriptions. However, the reader should recall the purpose of conservation tasks and then consider the general age ranges at which area and internal volume conservation develop. At the very least, questions are raised when it becomes clear that a child is expected to measure a quantity that is not yet functional developmentally at a minimal logical level, as defined by conservation.

Communicating

The communicating process first appears at the first-grade level in SAPA in five of the 26 process lessons. Example lessons are described here in enough detail to ensure that the reader can comprehend the scope of this process and what "concepts" are used to allow the process to function.

Lesson 1: Identifying Objects

The teacher begins this lesson by reading a story to the class about a "miscommunication" that occurs between a brother and a sister. A discussion follows the story (led by the teacher) wherein the teacher emphasizes the importance of "precise" information. A game called "information please" is shown to the children to "make concrete" the nature of precise communication. A set of eight cubes, two of a certain color, is used in the game. One child turns his/her back while the other points to one cube. Information is given to one child by the other until the "correct" cube is known. Other objects in the classroom can be used as well, with one child silently identifying an object while the other child tries to discover the object.

Another lesson deals with constructing a bar graph to represent a certain number of objects. Graph paper with 1-centimeter squares are used by the children along with cubes of various colors to construct the graph. After initial work with stacking blocks in front of a piece of vertical graph paper, children are given stones, seeds, or other objects to use as something to be graphed. Graphic communication, as the reader can probably see, would tend to require a much higher level of conceptualization than the previous example of communicating. It is expected that children will be able to "read" bar graphs as a result of these and related experiences. Many of the measuring lessons are related in that children are expected to master certain length-area-volume measures.

Of the three remaining process lessons on communicating, one of them also uses graphing as a means of describing certain phenomena. Children are asked to observe two colliding objects in terms of speed, weight, and distance moved. The teacher tells the children to observe the collision of rolling balls with a box and to determine which ball moves the box the greatest distance. Children are asked which ball exerted a greater force and then they are told to graph the results, using coordinate paper and pencil.

The reader is encouraged to determine the conceptual requirements of the lesson with the generally expected conceptual abilities of first-grade children.

Science Processes at the Third-Grade Level (SAPA)

The other two "basic" processes of science are introduced at the second-grade level and are continued during third grade. Inferring and predicting are among the 22 process lessons, although only a relatively small number of the total (6 of 22) are devoted to these somewhat more advanced processes. As before, a few example lessons for each process are briefly described to help communicate the nature of the progression of the science processes.

Observing

Three of the 22 process lessons are devoted to the process of observing. All of these lessons are considerably more advanced than the observing lessons found in the first-grade level. One lesson deals with magnetic poles, a second with uniform plant growth, and a third with falling times of objects. It is the third lesson that is described here.

The eight- and nine-year-old children in third grade are expected to describe and diagram the vectors of forces acting on falling bodies. Various objects are dropped and children are asked to observe and deduce such things as relative speeds, changes in speeds, and time of fall. Then groups of children are given some objects to determine relative falling speeds and upward and downward forces acting on the objects.

This lesson certainly raises some questions regarding the general conceptual level of most of the eight- and nine-year-old children as compared to the level required to cope in a meaningful way with this process lesson. Although identifying and controlling variables is not yet officially a part of the science processes for third grade, it is obvious that such thinking is required for a sensible interpretation of the experiment. Weight, size, shape, and position are all variables that can affect the falling times of objects and the experimenter must be able to separate these variables and then vary them systematically to make valid conclusions. It was clearly described in Part Two that eight- and nine-year-old children are not capable developmentally of such sophisticated mental gymnastics. Comparing relative speeds is not usually a functional ability until after the age of ten or eleven years and even then it is for more easily observable motions. If an object (ball, marble), is dropped from a height of 4 feet, it will reach the ground in only half a second! This means that a large experimental error is built into the falling bodies experiment even for formal operational types. With all of these problems, it seems safe to conclude that for most children in third grade, an experiment with falling bodies, where forces, relative speeds, relative times, and the like are involved, is conceptually above their observable developmental levels.

Using Space/Time Relations

Two of the three lessons on S/T relations involve relative position and motion. The rate of change of position is emphasized with two cars traveling given distances over various times. So the variables are speeds, distances, and times, and children are expected to "deduce" a general rule that states that speed is equal to distance moved per unit time. Various races among the children are also suggested for the teacher to help them understand the concept of rate of change of position (speed).

The reader is again referred to Part Two in the section on time and motion to aid in determining the "conceptual fit" of the current space/time lesson under consideration.

Using Numbers

Two of the 22 process lessons are directed specifically toward the use of (1) dividing to find speed, given distance and time, and (2) learning decimal names used in the metric system. No details are given here about the two lessons except to say that they are similar to fairly straightforward lessons commonly found in math texts.

Classifying

The only lesson identified as classifying in third grade uses a punch card system for recording a multistage classification system. Figure 4–13

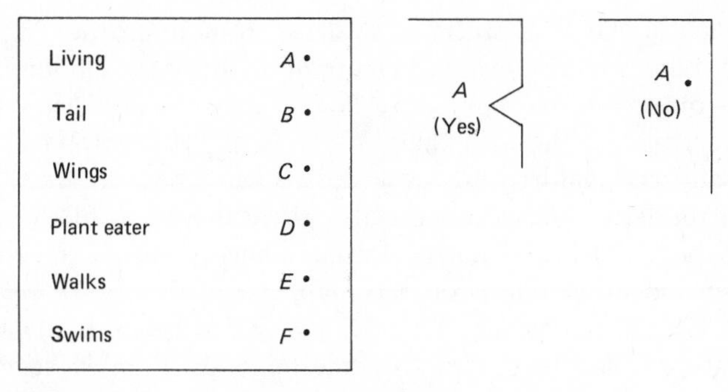

Figure 4–13. Punch-card classifying systems.

shows a type of punch card that can be used by children to record data about single objects or classes of objects.

As many attributes as desired can be used, and the child is given a task to classify the members of the class. Of course, many other attributes or characteristics can be used depending on the desired use.

Measuring

The four measuring process lessons are briefly these:

1. Comparing linear and circular motion in terms of distances. Measurements of rates of revolution of a disc are compared at various points on the disc and the distances traveled are compared.
2. Comparing "drops" of liquids as units of measure. Volumes of liquids are compared using a drop as the unit measure. The drop size can be varied by changing the size of the opening of the dropper or by varying the liquid.
3. Finding the rate of change in volume caused by evaporation. The rate of change is demonstrated in terms of units of weight, volume, or time.
4. Describing forces and motion with vector diagrams. Children are expected to identify forces causing motion and the effects of various forces.

Because of the rather advanced conceptual level of these measuring activities, it is again reasonable to raise the question of how the expected development levels of eight- and nine-year-olds coincide with the level of difficulty of the measuring lessons.

Communicating

Using maps is the first process lesson on communicating and involves scaling, naming locations, and changing map scales. Children are asked to draw objects in the room according to some "scale." After some work with identifying objects in their maps, "real" maps of their state are supplied and the children are expected to learn how to "read" the map by using the letter and number scales along the edges of the map. Coordinating the vertical and horizontal components allows the reader to locate cities and other places on the map.

At this point, the reader may be asking why this lesson is identified as communicating rather than as space/time relations. Frankly, it would be difficult to argue against renaming the lesson since it does seem to include mostly the use of a coordinate system and scaling systems (measuring). Even classification is directly involved since a kind of hierarchical classification is used with town or city, county, state, or country, comprising the various subgroups. The likelihood of eight- and nine-year-old children conceptualizing the process of mapping either in a qualitative or a quantitative sense is fairly small. A set of procedures for "reading" maps can be learned, but the corresponding meaning associated with the procedures is open to serious question, since the developmental levels of most third-grade children are not advanced enough to allow for generalization.

Inferring

The process of inferring is rather different than the other six processes described thus far. To make an *inference* one must use existing observations, measurements, and the like and then devise an explanation that is consistent with the existing data. The logical consistency of the inference to the data is directly related to the developmental level of the "inference-maker." Of course, the level of logic required will depend on the nature of the data (that is, concrete to abstract) and the nature of question-asking for the inference.

The first couple of inference lessons use mostly cartoons and pictures to help children distinguish between observations and inferences. By the third lesson the children are expected to actually construct inferences using a water-air system as in Figure 4–14. The teacher first demonstrates the system by blowing air into the bottle filled with water, using a drinking straw and then groups of about five children each are encouraged to do the same thing. The teacher then asks what they think happened to the water in the bottle (where did it go?). As inferences (explanations) are given, children are asked how they could test some of the ideas. Other demonstrations using similar apparatuses are used by the teacher to encourage children to make inferences about how things happen.

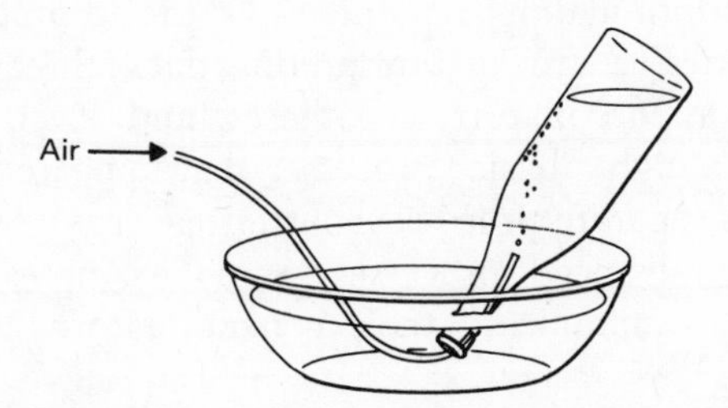

Figure 4–14. Air-water system for making inferences.

Predicting

Another process, like inferring, which is somewhat abstract by definition, is predicting. Based on a pattern of events, it is possible to predict the next occurrence *if* the pattern is understood. Being able to predict accurately the next event without actually trying it is, of course, an indication that an individual has been able to organize previous data into a recognizable pattern. It usually requires that one set of data be compared to another set in some ordered way.

One of the predicting lessons relates the bounce of a ball (height) to the dropping point (height). Also, the number of bounces can be related to the falling distance. Four groups of eight children are each given four different balls and told to record the number of bounces each one makes when dropped from a given height. The data are then graphed and discussed. Another activity involves similar materials, but this time the height of the bounce is compared to the falling distance. Data are again graphed and discussed and the teacher then asks the children to predict how high a given ball would bounce if it were dropped from x height. Children are expected to predict the bounce height by interpolating from the graph of the data already accumulated.

Predictions can vary greatly in terms of difficulty, as can any other process. A very easy prediction for the eight- or nine-year-old could involve simple expectations based on past experiences, with little conscious effort expended toward searching for a pattern. Just how closely the predicting lesson comes to the developmental level of most third graders is open to question. Qualitative predictions based on concrete experiences with bouncing balls is at quite a different level of difficulty than interpolating from data in graphic form.

Science Processes at the Fifth-Grade Level (SAPA)

Four "integrated processes" appear at this level and, in fact, at the previous level in fourth grade: controlling variables, defining operationally, interpreting data, and formulating hypotheses. Of the 18 process lessons for fifth-grade children, five are on interpreting data, three on defining operationally, three on formulating hypotheses, and seven on controlling variables. All of the eight "basic" processes that appeared at earlier levels no longer exist as separate processes, but instead have been "integrated" into these more advanced processes. Some examples of each process should help to communicate their general nature as a science process for children.

Controlling Variables

It was seen in Part Two that separating and controlling variables require a level of developmental ability that few children younger than about twelve or thirteen years of age seem to possess. As a developmental characteristic, controlling variables grows logically out of, first, being able to consider all possible combinations of factors (variables) that might have some impact on the outcome being studied and, second, isolating variables or their combinations one at a time. Investigations in science usually lead to "conclusion" statements about cause and effect. The purpose of controlling all variables except one allows the observer to draw certain logical conclusions about the cause(s) of certain effects. The implications of *developmental theory* for *learning outcomes* are discussed in more detail following a description of some example SAPA advanced or "integrated" processes.

The first lesson on controlling variables in the fifth-grade SAPA curriculum involves identifying and controlling variables that affect chemical reactions. Two photochemical reactions are the systems available for study: (1) blueprint paper, and (2) Ozalid paper. When part of a sheet of blueprint paper is exposed to light and the sheet is then immersed in water for a short time, the exposed area turns dark blue and the unexposed area turns white (chemicals merely wash off and expose the white paper underneath). The Ozalid paper will turn white from light exposure whereas unexposed areas will, when developed with an ammonia solution, turn blue, just the opposite of the blueprint paper. Exposure time, brightness of the light, and developing time are all obvious variables that could affect the shades of blue that result after the whole process. Of course, other less obvious variables might also affect the outcome. Such things as distance from the light source (which controls the brightness for artificial light sources), angle of the paper to the source, age of the paper (especially the chemicals on the paper), and the like must also be considered when conducting the investigation.

Children are shown by the teacher how the blueprint paper can be exposed and developed and are then instructed to find out how to do a similar thing. Then they are asked what variables could have caused differences in the shades of blue that result from the first experiment. Groups of six or seven students each then try to plan a procedure for testing one of the variables they think might account for the different shades of blue. Eventually all results are combined and discussed. A similar procedure is used with the Ozalid paper except this time the entire class suggests variables that could affect the outcome (shade of blue) and the groups of children are supposed to test one of the variables.

After the experiences with identifying and controlling variables, it is expected that the children will be able to design and conduct a systematic experiment involving a number of variables.

Another lesson on controlling variables uses small, warm-blooded animals, such as mice or gerbils, to "practice" the process of dealing in a systematic way with the control of variables. Children are encouraged to keep records of the various behaviors and characteristics of the animals, one of which is a male and the other a female. It is suggested that the female be weighed every third day to determine the rate of increase in weight as pregnancy (hopefully) progresses. After the young are born, the children are to keep records of all noticeable characteristics of the babies as they develop. About three weeks after the birth, the young animals are placed in a separate cage and the teacher asks, "What might be the result if each animal were fed a different diet?" When the diets are decided upon, records of weight, size, coat, and other variables are kept to see if diet has any observable effects on the young animals during a period of about 30 days.

On the surface this may seem like a fairly simple, straightforward experiment, but further reflection reveals many "hidden" variables that could have a significant impact on the observable outcomes. When young gerbils, mice, and other test animals are raised on a *similar* diet, there are usually observable variations that are simply the result of genetically determined physiological growth. This hidden factor could be controlled for but only through carefully designed experiments, not a very likely possibility with ten- and eleven-year-old children. Another complicating factor could be the sex of the members of each of the two groups of young animals, since the males of many species tend to be larger than the females. The amount of handling and possible injury to some of the animals caused by such handling by the children could well be another factor influencing their weight and general appearance. Other variables could also be identified that might have a significant effect on the observable characteristics, some of which would probably be rather difficult to pursue.

The point in all of this is that the process of identifying and controlling variables is more often than not a fairly complex one that requires a cognitive level that would be available to very few ten- and eleven-year-old children. Expectations of children should be in line with their capabilities rather than with disciplinary structure that has been developed over a long period of time by scientists capable of abstract, logical thought.

Interpreting Data

Both this science process and the formulating hypotheses process that follows suggest a "thinking approach," and the quality of the "product" should be dependent directly upon the developmental level of the person involved. As a process of science, interpreting data is something that must occur if the investigator seeks to explain or make sense out of natural phenomena. Collecting the data is certainly an important part of science processes, but a mere collection of data is incomplete unless some kind of explanation or interpretation, however tentative, accompanies the observations. Taking data and trying to see patterns that make sense has been interpreted by the developers of SAPA to mean two things: (1) interpreting data, and (2) formulating hypotheses. Examples of the first process, interpreting data, as intended for children in fifth grade, are now described.

One of the five lessons dealing with the interpreting data process is entitled "magnetic fields—the nature of the earth." Children are given a magnetic compass, a bar magnet, and some newsprint for mapping out and drawing the magnetic field around the magnet. After the drawings are completed, the teacher compares various drawings and directs the attention of the children to variations among some drawings and asks for explanations for them. The orientation of the magnet with respect to the earth's magnetic lines of force is a factor that will affect the nature of the children's drawings. Other activities with mapping the field around magnets are done and the children are then given some sketches of compass direction lines around a bar magnet and from these they are expected to determine the pole (see Figure 4–15). Maps of magnetic compass directions are shown to the children and they are asked if the North Magnetic Pole is a *N*-pole or a S-pole. It is intended that the children should be able to infer that it is a S-pole since the *N*-pole of the compass points toward it.

A second example of a lesson on interpreting data involves the naming of the probability of favorable and possible outcomes where many objects or events are a part of the system to be studied. The odds of something happening can be expressed by taking into account a theoretical probability and then making modifications to fit the empirical reality. In this lesson, the teacher explains the use of various types of symbolism and then the children are paired up to toss coins and record the numbers of heads and tails. All the groups then compare their results and discuss the variations. After the coin-tossing activities, the teacher leads a discussion on inheritance and probability using

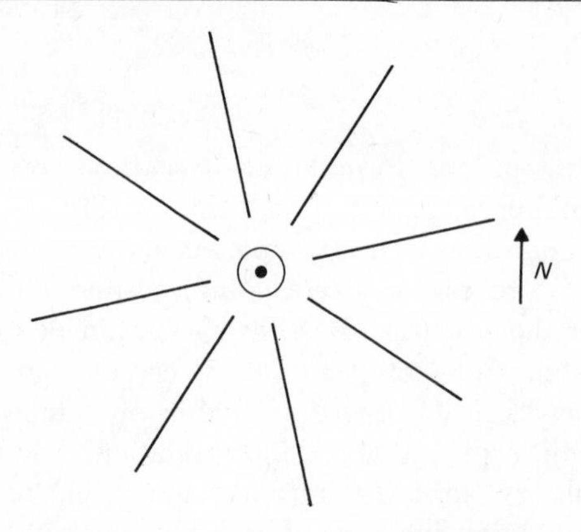

Figure 4–15. Compass direction lines around one pole of a vertical bar magnet.

"tasters" and "nontasters" as examples of dominant and recessive genes. A diagram as in Figure 4–16 is used to explain the probability of offspring being tasters or nontasters, given the genetic makeup of the parents. The diagram is supposed to help explain how the four possible outcomes between a *Tt* father and a *Tt* mother are *TT*, *tT*, *Tt*, and *tt*. Data sheets with other parental characteristics are used with the children to show how other combinations for the four possibilities are derived. Other activities intended to illustrate probability or possible outcomes include (1) males to females ratios in populations; (2) number of objects (beans, salt crystals) in a certain volume; (3) possible outcomes in a game of chance.

An interesting question to raise in thinking about the preceding lesson for children in fifth grade is, "What is the probability of their understanding the concept of probability?" The level of difficulty of the content as described in the teacher's guide seems to be fairly high, with the children being expected to determine quantitative probabilities from data that are often rather abstract (genetic inheritance).

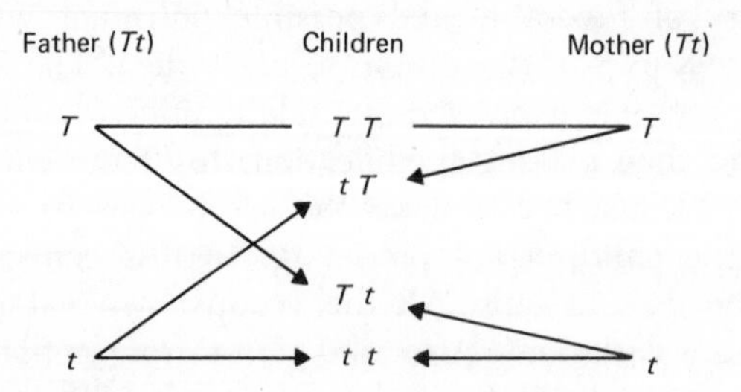

Figure 4–16. An inheritance diagram for tasters (*T*) and non-tasters (*t*).

Formulating Hypotheses

The process of making hypotheses based on results of experiments is a kind of generalizing process that also could be called "making up explanations." Hypothesizing, as the term implies, involves statements about hypothetical possibilities that are derived from some type of data base. The three lessons on formulating hypotheses that are a part of the SAPA program in fifth grade require that the students construct hypotheses from data and, conversely, identify data that support or do not support a hypothesis made by the student.

An example of such a lesson deals with the effects of temperature on dissolving rate of several substances. The teacher begins the lesson by placing two clear containers of water, one cool and the other hot, where the students can see them and then places a sugar cube in each container and stirs until the sugar cubes are dissolved. Since one cube dissolves before the other, the students are asked why, and some will usually say one of the containers is warmer than the other. The teacher then asks a group of six students to repeat the demonstration using a watch with a second hand to measure how long it takes for each cube to dissolve. After this "experiment," three or four groups of students are formed to check on the effects of different water temperatures on the rate of the dissolving of seltzer tablets. At the same time they are reminded that other variables could have some effect on the outcome must be controlled. Each group is assigned a specific temperature at which the test should be run and that the children are to check their results by doing it a second time. After the data (dissolving times) are collected, the students are told to find the average or mean time and record the data on the chalkboard. They are also directed to plot the dissolving time and temperature on graph paper and then predict what the dissolving times for various temperatures might be by interpolation and extrapolation. Further testing of the predictions occur and the children find (perhaps) that the dissolving time at various temperatures is not a linear function. That is, the line graph is not straight but, rather, curved as in Figure 4–17.

As can be seen from the brief description of this lesson on formulating hypotheses, many other processes such as interpreting data, predicting, and measuring are involved. It was noted earlier in Part Two that one important characteristic of formal or logical thought is the ability to hypothesize about possibilities. Making generalizing hypotheses based on a logical analysis of data could not be expected of ten- and eleven-year-old children if cognitive development research and theory were to be taken seriously, *unless* learning is defined merely as rote memory of

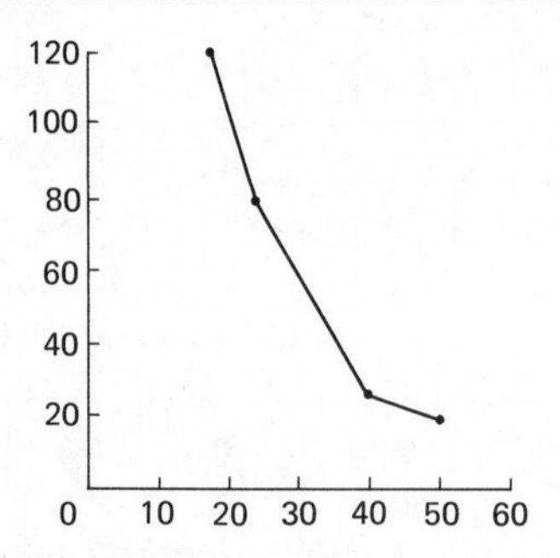

Figure 4–17. Dissolving time graph for seltzer tablets.

information and procedures. Teaching youngsters how to find the second derivative of a mathematical function, as in a calculus course, is a far different thing than trying to teach them why the procedures are used and where they come from.

Piaget has noted that development explains learning. The nature of what can be learned is limited by the cognitive development level of the learner. Genuine learning, which is stable and generalizable, cannot exceed the operational schemas available.

When something is defined operationally, it usually means that it is defined in terms of "how something works." One lesson that illustrates operational definitions involves true north and magnetic declination. Children are shown various techniques that can be used to represent magnetic declination (the angle east or west of true north) and are then asked to show how they would find the values of magnetic declinations of various geographic points.

Another lesson shows the children various examples of mass and inertia with moving objects and then asks them to construct an operational definition of mass using an equal-arm balance or some other device.

In general, the operational definition refers to a kind of demonstration that illustrates the physical meaning of a term.

Summary of Science Processes and Children's Thinking

When science is defined as a process approach, the content or concepts are chosen merely to allow the processes (observing, classifying, measuring) to function. By definition, a process approach to science implies an active, "hands-on" role for the students. It does not, however, imply just what the teacher's role is to be or *how* the students are to engage in the process approach. The teacher's role, as seen in Part Five, is very important and cannot really be separated from the science curriculum.

The conceptual level of a curriculum based upon a set of science processes can vary just as much in level of difficulty as one based upon science concepts. Since there was a considerable amount of detail given to conceptual level of science concepts in the first section of Part Four, it was assumed that the reader could just reapply the methods in this latter section on science processes. The most fundamental flaw in most of what has been described as science concepts and processes for children thus far is the implicit assumption that all children at a given grade level are equally capable of understanding what is going on. One thing that is completely apparent from the research and theory on cognitive development is the 2- to 3-year range of abilities among children at any given grade level. If one assumes that learning with understanding (conceptual learning), as opposed to rote or skill learning, is dependent upon the developmental level of the learner, then a curriculum that reflects a single level of difficulty for all children must be deemed as conceptually inappropriate.

Most curriculum materials are highly subject to alteration by a teacher. In fact, there are many studies that identify the teacher as the single most important variable in affecting student behaviors and outcomes. We now turn our attention to this variable.

part five

THE ROLE OF THE TEACHER

objectives

It is intended that this section will help the reader to

1. Identify important conditions that a teacher should consider in helping students learn science.
2. Identify and describe five teacher roles that can be used in teaching science.
3. Compare the teacher role of *lecturer* with three learning conditions that grow out of our current understanding of the intellectual development of children and with the nature of science.
4. Compare the teacher role of *discussion leader* with three learning conditions and the nature of science.
5. Compare the teacher role of *activity director* with three learning conditions and the nature of science.
6. Compare the teacher role of *activity facilitator* with three learning conditions and the nature of science.
7. Compare the teacher role of *co-investigator* with three learning conditions and the nature of science.
8. Determine the nature of *quantitative* studies of teacher and student behaviors in the classroom learning environment.
9. Analyze example science texts for students in terms of the five teacher roles identified earlier.
10. Analyze example science curriculum projects (SAPA, ESS, SCIS) in terms of the five teacher roles.

introduction

Thus far we have established that children develop certain intellectual capabilities in a generally predictable sequence, although the rates of development vary among children. One child might be assessed and identified as a conserver of number, whereas another child of a similar age is a nonconserver of number. In fact, in a given classroom of children whose chronological ages vary by an average of only a few months, it is usually found that cognitive development levels vary by 2 to 3 years. If it is assumed that conceptual learning is limited by one's developmental level, then what can be conceptualized varies widely among the children in almost any given classroom. This is identified as one of four major problems or conditions facing a teacher in helping children "learn" science.

CONDITION ONE: **There are wide variations in developmental levels among children in most classrooms and, thus, equally wide variations in their conceptualizing abilities.**

Another condition that faces the teacher who wants to help children learn science grows out of the *theory* as compared to the *results* of cognitive development research. Although this book has emphasized Piaget's experiments much more than his theory, it is the latter that has won Piaget so much acclaim in developmental psychology, science and mathematics education, and many other areas of study. The central part of his theory is that of how the child adapts to the environment to develop increasing powers of cognition and conceptualization. The very personalized, active process of cognitive growth is termed *equilibration* by Piaget. Although the nature of this process changes continually as the cognitive abilities develop in sophistication (toward hypothetico-deductive reasoning), most of the domain of the process requires a physical interaction by the child with the environment.

CONDITION TWO: **Conceptual learning is tied closely to developmental level and occurs primarily as a result of a child's internalizing his or her own actions on objects.**

This "condition" of learning defines the general nature of the learning environment *if* conceptual learning rather than skill learning is the desired outcome.

Another major condition that must be considered by the science teacher grows out of a study of the field of cognitive development and related research. Specifically, this third condition concerns the susceptibility of children's developmental levels to direct training. Scores of studies involving attempts to accelerate conservation and other cognitive indicators have been reported and include: Smedslund (1961a, 1961b, 1961c, 1961c, 1961d, 1963), Wallach et al. (1967), Gruen (1965), Beilin (1962), Murray (1968), Brainerd and Allen (1971), Almy (1970), Leskow and Smock (1970), Emery (1973), Good et al. (1974), and many others. Most of the studies reporting attempts at accelerating cognitive levels have met with very little or limited success. Those that have reported large gains usually report little, if any, transfer involved in the apparent learning. Although the issue of acceleration is far from being settled at this point, it seems fair to say that Piaget's original contention about the resistance of development to "artificial" means, is still intact. This brings us to the third condition.

CONDITION THREE: **It appears that direct, verbal instruction has little effect in facilitating advancement of developmental levels and, thus, the ability of children to conceptualize problems, beyond their current developmental abilities.**

The three preceding conditions, which grow out of the nature of children's thinking, lead us to look at the role of the teacher in helping children learn science. It is generally agreed upon by those who study the teaching–learning process in our schools, that the teacher is the single most important variable in determining the general classroom environment. Keeping in mind the three conditions growing out of cognitive development theory and research, we now study the teacher's roles in determining how children learn science.

teacher roles

Our conception of "teaching" is usually arrived at through a process of synthesizing past experiences with teachers. The resultant composite model of teaching is most often that of the knower trying to communicate the known to a student who usually does not voluntarily choose to learn the known.

We consider here five distinctly different roles that can be assumed by the teacher who wants to help children learn science. Before doing this, however, we must arrive at a definition of science.

Defining Science

A rather simplistic way of defining science, or any other area of study for that matter, is to say it is "what scientists do." Art becomes "what artists do." History becomes "what historians do," and so on. In effect, this means that the nature of any discipline is determined by those who have been and currently are engaged in the process of making up the ideas. A current definition of science would be determined by those persons currently engaged in the discipline.

During the early 1960s the largest organization of scientists in this country, The American Association for the Advancement of Science (AAAS), sanctioned a Commission on Science Education to develop a science program for children that would more accurately reflect the true nature of science. This true nature of science was determined to be *process-oriented.* Children would engage in the *processes* of science (which have been outlined in Part Four) rather than the *products* (knowledge outcomes). De-emphasizing the content of science (facts, concepts, principles) in favor of processes represented a radical departure to science for children, particularly when it is considered that all science textbooks for children were replaced with manipulative materials. Words were to be replaced with actions.

When science is viewed as *an active process of manipulating one's environment in order to find out more about it,* there are many important implications for science teaching. Learning science under this definition would be very different than learning science as an accumulation of previously identified facts, concepts, and principles. Any consideration of various roles for the teacher who is trying to help children learn science must include a clear definition of science. This "condition" is added to the previous three conditions that evolved from our study of the nature of children's thinking. All four conditions are stated here, prior to considering possible roles that could be assumed by the teacher who wants to help children learn science.

CONDITION ONE: **There are wide variations in development levels among children in most classrooms and, thus, equally wide variations in their conceptualizing abilities.**

CONDITION TWO: **Conceptual learning is tied closely to developmental level and occurs primarily as a result of a child's internalizing his or her own actions on objects.**

CONDITION THREE: **It appears that direct, verbal instruction has little effect in facilitating advancements of developmental levels and, thus, the ability of children to conceptualize problems, beyond their current developmental abilities.**

CONDITION FOUR: **Science is an attempt to make more sense out of one's environment by manipulating portions of that environment at a difficulty level consistent with one's conceptual abilities.**

Constant references are made to these four conditions in assessing the "appropriateness" of various teacher roles, so the reader should be able to relate each condition with the corresponding numeral.

Five Teacher Roles

Five distinctly different teacher roles or instructional strategies have been chosen to study in some detail, relative to the four conditions just identified. These five roles are:

A. Lecturer and Question-Answerer.

B. Discussion Leader.

C. Activity Director and Evaluator.

D. Activity Facilitator.

E. Co-investigator.

Roles A and B are primarily verbal in nature and the corresponding learning on the part of the children is, by definition, largely verbal also. *Role C* implies that children will be engaged in "activities," and the teacher will direct and evaluate their activities. The learning here will be a combination of verbal (from directions and evaluations by the teacher) and nonverbal (from the child's own actions during activities). In *Role D* the teacher avoids evaluating and directing behaviors in favor of accepting and nondirective behaviors. Because of this shift in teacher behaviors, nonverbal, self-directed learning opportunities increase while verbal learning decreases. As "co-investigator" in *Role E,* the teacher engages in genuine investigative activity along with the children. Verbal interaction here is usually initiated by the child rather than by the teacher.

Each of these five teacher roles now is studied in more detail.

Role A: Lecturer and Question-Answerer

The best concrete example of this teacher role is found in most large enrollment (50+) courses in college. In this role, the teacher devotes all or almost all of his or her time to information-giving while students devote their lesson-related time to recording the information in one form or another. This teacher role is not exclusively restricted to postsecondary education, however. Although the classes are smaller in our secondary school classrooms, much of what the teacher does still falls under teacher Role A. At the elementary school level Role A becomes far less common.

The four conditions or considerations that have been identified as important in determining how children learn science can and will be used with each of the five teacher roles.

Role A and Condition One: Incompatible

It should be fairly obvious that any "single verbal presentation" type of teaching role is unable to take into account the variations in conceptual abilities of the "audience."

Role A and Condition Two: Incompatible

Since conceptual learning is heavily dependent upon children's taking a "hands on" approach, the passive audience of the lecturer has greatly reduced opportunities for such learning. Learning science for children in such a passive situation is reduced largely to learning words that may make little sense to them.

Role A and Condition Three: Incompatible

The fact that a lecture role for the teacher *is,* by definition, clearly a verbal role means that Role A will have little effect in facilitating cognitive development and thus conceptual learning in children.

Role A and Condition Four: Incompatible

Active manipulation of one's environment is not very probable for the student who sits listening to a barrage of terms, unless that student chooses to engage in "nonlesson-related behavior" by watching a fly walking along the desk top.

For Role A, all four conditions for conceptual learning clearly are not satisfied. There are certainly others that merit consideration such as interest and attention span, but the scope of this section remains limited to the four conditions that have been defined. The personality and general emotional characteristics of the teacher are certainly critical components of the teaching-learning process as well, and the lack of reference to the affective domain in this book should not be interpreted as a slight to the importance of this domain. It is generally agreed upon by practically everyone in education that a teacher of children should be a warm, caring, sensitive, patient, open, honest, and dedicated person. *Feelings* such as these are extremely difficult to teach, however, and the scope of this book does not allow for the space that would be required for even a minimal treatment of the subject. Thus, it will be assumed that such positive, humane characteristics are *prerequisites* to learning how to help children learn science.

Role B: Discussion Leader

Where student groups are not too large, it is possible for the teacher to assume the role of discussion leader. Most of us have been in classes where "discussions" are going on and know that usually fewer than one-half of the students contribute in any significant way to the discussion. In elementary school classrooms, particularly at the earlier grade levels (K–2), the egocentric nature of the children's thought drastically reduces the likelihood of conceptual learning occurring during discussions.

Role B and Condition One: Partially Compatible

A discussion at least allows for verbal participation on the part of the students, whereas the lecture role minimizes their participation. As discussion leader, the teacher can guide the direction of the experience and allow, to some degree, for the variations in developmental abilities of the children. For this reason, Role *B* was identified as "partially compatible" with conceptual variations that are sure to occur and be exhibited through the nature of children's explanations, questions, and other responses.

Role B and Condition Two: Incompatible

Verbal interaction without corresponding "actions on objects" by children has minimal positive effects on the advancement of children's thinking abilities. A discussion in science that requires any thinking naturally assumes that the participants are capable of conceptualizing various natural phenomena (remember some of the examples from Part Four) *without the actual physical experiences.* Or, perhaps the discussion might occur after some "common" experience(s) (common in the physical sense only since Condition One holds) by the children. This tends to promote "closure" and social interaction skills but does very little to facilitate further conceptualization on the part of the children. Discussions, of course, can vary widely according to how much "control" the teacher exerts over the group. The teacher can be positive to negative, open to closed, accepting to rejecting, or the like in reacting to the children's comments and the actual composite behavior is very complex indeed. Discussion leader *as the main role* of the science teacher, however, must be viewed as incompatible with Condition Two.

Role B and Condition Three: Incompatible

In most cases Conditions Two and Three are directly related in that a predominance of one, say verbal instruction, implies a lack of the other. Since discussion is usually a form of direct, verbal instruction or indirect, verbal instruction, little effect in conceptual learning can be expected.

Role B and Condition Four: Incompatible

The discussion leader role hardly facilitates a learning environment in which children can physically manipulate portions of their environment and, thus, engage in actions that will lead to developmental and conceptual growth.

Role C: Activity Director and Evaluator

As the name implies, the teacher is involved with a science curriculum that consists of manipulative materials and, possibly, some written materials. Children are engaged in "hands-on" activities. The role of the teacher is to direct the children's activities and evaluate their progress according to product or process measures. Roles *A* and *B* generally are associated with only the learning of facts and concepts (products), whereas Role *C* can be associated with *either* process or product outcomes. The teacher in this role can direct the students' activities toward specified facts that are a part of their experiences, or attention can be focused on processes, as in Science: A Process Approach (SAPA). In either case, it is the teacher who directs the activities of the children toward prespecified learning and then evaluates in some fairly straightforward way to determine the success of the instruction. The evaluation could be in the form of immediate feedback to the student during the activity (for example, "that's right," "that's wrong," "that's good," "that's close") or immediately after the lesson or both. Again, using SAPA as an example, the teacher is expected to evaluate and give immediate feedback to the students as well as evaluate what was learned from the overall lesson.

It is possible for the teacher assuming Role *C* to "individualize" the activities to the extent that various children (or groups) can be directed to do different activities or variations within the same general activity. The reader might remember that many of the example lessons in SAPA described in Part Four did, in fact, suggest that the teacher direct groups of children to work on various aspects of the overall lesson. Summarizing experiences were then provided by the teacher in an attempt to ensure that all of the children received approximately the same learning opportunities.

Role C and Condition One: Partially Compatible

The fact that Role *C* emphasizes teacher directiveness and evaluation does not really explain how well Condition One is accounted for. If all children are directed to do the same thing and evaluated according to similar criteria, then one could hardly say that the variations in conceptual abilities of the children are taken into consideration. On the other hand, if there would be some means of assessing the cognitive characteristics of each child related to the activity in question, then it might be possible to allow for Condition One, at least in a general sense, by matching up expectations with abilities. Except for fairly simple skills, however, it is not really possible for the classroom teacher to carry out the time-consuming assessments with 30 or so children, which would be required for each different activity. This means that even under the best conditions, teacher Role *C* can be only partially compatible with Condition One.

Role C and Condition Two: Partially Compatible

The fact that the science curriculum consists of "hands-on" activities by children for a significant part of the time means that some of their physical actions can be internalized as operations. A heavy emphasis on directions by the teacher, however, tends to promote a kind of imitation on the part of children who are unable to conceptualize what the teacher intends and this inhibits the structuring of real operations. For this reason, even if a science curriculum is centered entirely around manipulative materials, as in SAPA, Role *C* is only partially compatible with Condition Two.

Role C and Condition Three: Partially Compatible

Role *C* involves direct, verbal instruction by the teacher and is in conflict with earlier information about the relative ineffectiveness of such a strategy with children, if conceptual learning is a desired outcome. Since children will have some opportunities for self-structured activities with manipulative materials, Role *C* is considered partially compatible with Condition Three.

Role C and Condition Four: Partially Compatible

A teacher who assumes Role *C* in helping children learn science most certainly will direct many children in a given class to engage in conceptual activities that are beyond their developmental abilities. A child who is directed to conceptualize content or processes beyond his or her current ability will not be able to "do science" as defined by Condition Four.

Role D: Activity Facilitator

The word *facilitator* is used in the sense of one who assists without directing or evaluating the actions of those being helped. A teacher who assumes the role of facilitator consciously must avoid directing children into predetermined activities and instead encourage self-direction and divergent behavior of those in a given group. In this sense, it is possible for a discussion leader (Role *B*) to facilitate divergent verbal interactions among those in the group, but Role *D* assumes a science curriculum that is centered around mostly nonverbal activities using concrete systems. When the teacher in Role *D* asks a question, it is actually intended only to elicit a response from the child, whereas in Role *C*, a statement in question form is often intended to direct or evaluate the child. If classrooms are available to study the questioning behaviors of teachers, it is very interesting to determine what proportion of the questions are actually "disguised" directives. As in all other roles, it is assumed that the teacher will develop a nonthreatening, caring atmosphere where the children like the teacher as a person. It is important in Role *D*, however, to avoid evaluative statements that suggest "correctness" to the children. The facilitating role must be accepting, not evaluative. Especially for younger children, verbal praise can be an effective "directing" strategy which, of course, is appropriate for Role *C*, but not for Role *D*.

It can be argued that merely identifying the activity for the children is a kind of directive and that is true, if they have no other alternatives. However, as long as a child interprets the introduction of a new activity as one alternative in many and as long as directive and evaluative behaviors are generally avoided by the teacher when children are engaged in the activities, it is reasonable to assume that children will view the teacher's role as facilitator rather than director.

Role D and Condition One: Compatible

Since it is virtually impossible for the classroom teacher to make accurate assessments of cognitive development levels of the 30 or so children in the class, a more logical approach to the problem is to allow and encourage each child to engage in conceptual activity determined by his or her developmental level. This strategy requires activities that make it possible for children to "work into" the activity at various conceptual levels. The range of difficulty of possible activities for the children must be at least as wide as their range of developmental abilities. This requirement places great importance on choosing or developing "appropriate" science activities, examples of which are considered later in Part Five. For now, it is sufficient to say that Role *D* does allow for the wide variations in conceptual abilities that are likely to be found in any group of children.

Role D and Condition Two: Compatible

Since it is the child who determines the specific problems to be solved and the science program consists primarily of an "actions on objects" approach, conceptual learning is maximized. Evaluation is unnecessary because whatever lesson-related behaviors the child chooses to exhibit, the teacher must accept and encourage.

Role D and Condition Three: Compatible

Very little directive behavior can be a part of Role *D* and most of this usually comes during the introduction of a new activity. The large majority of the children's productive activity will be directed toward the materials at hand and in attempts to find out more about objects and events that have been identified by the learner as interesting to pursue. The teacher interacts with individuals and can question, challenge, and so on as long as the child interprets the statements and gestures as alternatives rather than directives.

Role D and Condition Four: Compatible

Since science has been defined as "attempts to make more sense out of one's environment by manipulating portions of that environment at a difficulty level consistent with one's conceptual abilities," it should be clear that Role *D* satisfies both parts of this condition. As long as a child engages in self-direction by choosing from among appropriate alternatives and evaluates his or her own progress during the investigative process, the experiences will make sense *at the child's level.*

Role E: Co-investigator

The fifth and final teacher role to be considered here usually occurs at the graduate research level in universities. A senior researcher or faculty advisor works with graduate students or postdoctoral fellows as "co-investigators" in research projects (although most of the credit is usually taken by the "senior" person). The word *co-investigator* suggests that both parties genuinely are involved in trying to find out more about similar or at least related phenomena. This strategy has tremendous advantages because it tends to communicate a seriousness of purpose and a friendly respect between the co-investigators. Trying to translate this strategy into a workable role for the teacher who wants to help children learn science becomes more difficult, although quite feasible with younger children. Because of the egocentric nature of children's thinking until about eight or nine years of age, "co-investigative activity" remains somewhat dubious, unless the "researchers" are working on different problems. Under these conditions, the teacher conducts experiments with plants, insects, various physical systems, and so on, while students also conduct their own experiments using appropriate manipulative materials. If fourth-grade children are using microscopes to observe pond water, blood, salt, sugar, hair, and the like, the teacher might be using his or her microscope to conduct similar investigations. The children should view the teacher's role as co-investigator and, in fact, very little interaction would be initiated by the teacher, as in the previous four roles, for the purpose of questioning, encouraging, and evaluating after activities have been introduced. This does not mean that the teacher cannot spend significant amounts of time with children's experiments. It only means that Role *E* requires that for a significant amount of time the teacher conduct experiments at the same time that children are engaged in their own science activities. When interactions between teacher and students do occur, the teacher role is similar to that already described as Role *D*.

Role E and Condition One: Compatible

Just as in Role *D*, the teacher who assumes Role *E* allows for the variations in conceptual abilities by encouraging each student to identify problems of interest and seek solutions. In Role E, the teacher pays more attention to his or her own experiments and encourages the students to do the same.

Role E and Condition Two: Compatible

The primary mode of learning is still through an "actions on objects" approach and so conceptual learning is facilitated by these physical conditions.

Role E and Condition Three: Compatible

Much less verbal interaction occurs for this teacher role than for Role Four and so the students have even fewer opportunities to interpret the teacher's "verbalisms" as "directives." The teacher's intents in both cases are to encourage without directing and communicate an acceptance without evaluating.

Role E and Condition Four: Compatible

Since this teacher role is drawn primarily from the actual role of the scientist as co-investigator with colleagues, students, and others, it seems fair to say that Role *E* comes closer to our original definition of science than any of the other four roles.

Summary of the Five Teacher Roles

Five roles for the teacher who wants to help children learn science have been described:

A. Lecturer and Question-Answerer.

B. Discussion Leader.

C. Activity Director and Evaluator.

D. Activity Facilitator.

E. Co-investigator.

Each of the five roles has been compared with the three cognitive development conditions:

1. There are wide variations in developmental levels among children in most classrooms and, thus, equally wide variations in their conceptualizing abilities.
2. Conceptual learning is tied closely to developmental level and occurs primarily as a result of a child internalizing his or her own actions on objects.
3. It appears that direct, verbal instruction has little effect in facilitating advancements of developmental levels and, thus, the ability of children to conceptualize problems, beyond their current developmental abilities.

A fourth condition involving the nature of science was also identified and compared with each teacher role.

4. Science is an attempt to make more sense out of one's environment by manipulating portions of that environment at a difficulty level consistent with one's conceptual abilities.

Table 5–1 shows the "compatibility results" between each of the five teacher roles and the four conditions. The "*I-C*" is used to designate "partially compatible." It is readily seen that Roles *A* and *B* are very similar in terms of the four conditions, and the learning outcomes will be primarily verbal in nature.

Because Role *C* requires that children engage in "hands-on" activities, while the teacher directs and evaluates, this role is sort of midway between the others in terms of the four conditions and the nature of the learning outcomes. And finally, Roles *D* and *E* receive nearly the same "scores" indicating a science class where "hands-on" activities are occurring, whereas the teacher is nondirective and accepting.

A great many studies have been done on the teacher role in science

TABLE 5–1. **Five Teacher Roles Compared to the Four Learning Conditions**

Teacher Role	Learning Condition (Incompatible = *I*, Compatible = *C*) 1	2	3	4	Type of Learning (Verbal = *V*) (Nonverbal = *N*)
A	*I*	*I*	*I*	*I*	*V*
B	*I-C*	*I*	*I*	*I*	*V*
C	*I-C*	*I-C*	*I-C*	*I-C*	*V* & *N*
D	*C*	*C*	*C*	*C*	mostly *V*
E	*C*	*C*	*C*	*C*	*N*

classes, but nearly all of them suffer from a lack of specifying accurately (quantitatively) the conditions that were being studied. Adjectives are very unreliable when used in scientific research because there are so many different possible interpretations, depending upon a person's past experiences. "Open," "directive," "discovery-oriented," "traditional," and so on communicate a variety of meanings and for this reason, a description of learning conditions must be defined quantitatively. In particular, the teacher's role must be defined in terms of observable behaviors. This does not mean that the less quantifiable, affective part of a teacher's role (warmth, caring, sensitivity, patience, understanding) is not of high importance. It merely means that *where it is possible,* classroom conditions that are studied in a scientific sense should be quantified so that what is communicated by way of research results can be *clearly* interpreted. This author has been involved in such a study and brief descriptions of various phases of the project are contained in Appendix A. Most of the research was conducted in elementary science classrooms involving about 250 children and at the secondary school level in chemistry involving about 100 students. The reader is encouraged to take some time at this point and read the information in Appendix *A*. It should help to clarify the five teacher roles that have been described thus far and, also, provide a better perspective for the remaining sections in Part Five.

the teacher and the science curriculum

In Part Four, examples of commercially available science curricula were grouped into two general categories (1) those with *content* emphasis and (2) those with *process* emphasis. Most textbook series reflect the former whereas many curriculum "projects" such as Science: A Process Approach (SAPA) reflect a "process" approach by avoiding the use of a textbook as the mainstay of the curriculum and instead using manipulative materials. A "content" approach with the textbook as the mainstay of the curriculum tends to encourage "verbal" learning such as definitions and rules, which are displayed in written or pictorial form in the book. A "process" approach with manipulative materials as the mainstay of the curriculum tends to encourage "nonverbal" learning such as how parts of a system of manipulative materials fit together and how they interact.

The nature of the science curriculum (that is, the "things" or "materials," whether textbooks or manipulative materials) can have a direct influence on the type of teacher role that seems to "go along with" the curriculum. Some common curriculum "types" in science for children are described with corresponding teacher roles.

Textbooks and Teacher Roles

The nature of the content of science texts for children was considered in some detail in Part Four. Various means for analyzing the content were suggested and the reader was encouraged to find a textbook or two and try to make some "quantitative" assessment of the material between the covers. The role of the teacher is usually made explicit in the "teacher's edition," which accompanies all science series at both the elementary and secondary levels. Most textbook series suggest a teacher's role, which is a combination of lecturer and question-answerer (Role *A*) with that of discussion leader (Role *B*). Suggestions for the introduction of manipulative materials is often by way of demonstrations by the teacher. The role of the student is primarily that of listener (to teacher or peer talk), watcher (of teacher demonstrations), and reader. Student experiments, when they appear, are often unrealistic in terms of the equipment required.

Of course, no two textbook series are alike and some give more emphasis to student experiments than others. Overall, however, *most* science series for children give much more emphasis to learning science as content than as process. The reader is again encouraged to review a science text or two to assess the role of the teacher. One means of doing this, consistent with the five teacher roles previously described, would be to estimate the percentage of time the teacher is expected to assume the various roles. Example results are here illustrated:

Analysis of Text A

Role A—Lecturer and Question-Answerer: 40 per cent.
Role B—Discussion Leader: 40 per cent.
Role C—Activity Director and Evaluator: 10 per cent.
Role D—Activity Facilitator: 10 per cent.
Role E—Co-investigator: 0 per cent.

Analysis of Text B

Role A—Lecturer and Question-Answerer: 20 per cent.
Role B—Discussion Leader: 50 per cent.
Role C—Activity Director and Evaluator: 25 per cent.
Role D—Activity Facilitator: 5 per cent.
Role E—Co-investigator: 0 per cent.

It is quite likely that teacher roles not a part of *A–E* will turn up and in that case new categories can be added. Attempts at more quantitative predictions of teacher roles can be accomplished by using checklists for teacher behaviors such as the one found in Appendix *A*. Such attempts are more time-consuming but are also likely to be more reliable.

Curriculum Projects and Teacher Roles

Curriculum projects for elementary schools refer to a phenomenon of the 1960s, when the federal government, mostly through the National Science Foundation, poured large sums of money into attempts to improve science education. The three largest and best-known projects at the elementary school level are now described briefly in terms of the teacher roles. Each project provides for a different teacher role because of the variety of views of the developers regarding the nature of science, the nature of children, and the nature of "teaching" and "learning."

Science: A Process Approach (SAPA) and Teacher Roles[1]

This elementary school science curriculum is radically different than the textbook curricula in that no textbooks are a part of the materials for children. Most materials that were developed are manipulative in nature for the expressed purpose of getting children to engage in the "processes" of science (classifying, observing, measuring, and so on). The teacher role is primarily *Role C*, Activity Director and Evaluator. Chil-

[1] Further information on SAPA can be obtained from the publisher, Xerox.

dren are directed by the teacher as to *what* should be done with materials, *how* it should be done, and *if* they have performed the tasks correctly. There are variations of teacher roles within each grade level and from one level to the next, but the overall instructional strategy is very clear: "Children should be directed to engage in prespecified behaviors in order to build a repertoire of process skills that are progressively more difficult."

Since Part Four contains many example SAPA lessons, no more space is devoted here to a description of the program and the corresponding role of the teacher.

Science Curriculum Improvement Study (SCIS) and Teacher Roles

Another federally funded curriculum development project of the 1960s was the Science Curriculum Improvement Study (SCIS).[2] The program developers believed that science for children should be built around "conceptual schemes" and Figure 5–1 shows the general framework of the SCIS curriculum.

Descriptions of example activities within some of the units should help in assessing the general role of the teacher who implements the SCIS curriculum as described by the teacher's guides.

Organisms

This first grade unit uses seeds, plants, guppies, snails, daphnia, and other organisms to allow the children to observe and experiment with

FIGURE 5–1. **SCIS "conceptual schemes" framework.**

Grade	Physical Science Units	Life Science Units
1	Material Objects	Organisms
2	Interaction and Systems	Life Cycles
3	Subsystems and Variables	Populations
4	Relative Position and Motion	Environments
5	Energy Sources	Communities
6	Models: Electric and Magnetic Interaction	Ecosystems

[2] Further information on SCIS can be obtained from the publisher, Rand McNally.

"real things." Over a period of about 15 weeks the children are guided by the teacher to carry out the following activities, as well as others:

1. Observe and plant various seeds.
2. Construct and observe aquaria.
3. Observe the birth and growth of guppies in the aquaria.
4. Observe a dead organism in the aquaria.
5. Observe outside "habitats."
6. Observe algae in the aquaria.
7. Filter algae from the water.
8. Observe daphnia as a food source for the guppies.
9. Observe the decaying matter at the bottom of the aquaria.

The role of the teacher, as outlined in the teacher's guide, is divided among three of the five roles identified earlier:

Role B: Discussion Leader

Role C: Activity Director and Evaluator

Role D: Activity Facilitator

Children are often told what and how to do things (Role *C*), but the teacher is also expected to encourage a significant amount of open exploration (Role *D*). Discussions (Role *B*) usually follow activities by the children. The developers of SCIS outlined three stages in most lessons: (1) exploration, (2) invention, and (3) discovery. These words are actually a bit misleading because they all tend to suggest that it is the *child* who explores, invents, and discovers when, in reality, the teacher often invents and discovers for the children. Evaluation is not emphasized as it is in SAPA.

Interaction and Systems

This second-grade unit introduces the central concept of the entire SCIS program, interaction. Some of the activities outlined for the children in the teacher's guide are these:

1. Floating and sinking objects.
2. Grouping objects.
3. Matching properties with objects.
4. Making systems of batteries, bulbs, wire, paper clips, and other objects.

5. Using photographic paper.
6. Comparing air and breath.
7. Using heat-sensitive paper.
8. Constructing pulley systems.
9. Dissolving colored crystals in water.
10. Studying magnetic interaction.
11. Constructing electric circuits.

As in the Organisms Unit, the three teacher roles (*B*, *C*, *D*) are evident throughout, but Role *B* is emphasized more in this unit. Another difference is that the students now have a "manual" in which they record observations and respond to certain questions and directions by circling pictures, making drawings, and other techniques. This added component, of course, causes a corresponding change in the teacher's role, primarily by increasing Role *C* (activity director and evaluator).

Populations

Populations of small plants and animals are studied in this third-grade unit during a period of about 16 weeks. Populations of peas, crickets, aphids, flies, daphnia, and other organisms are studied in terms of interaction with the environment. The numbers of aphids, daphnia, and so on are emphasized by the teacher, and graphs are constructed to keep track of daily populations. The predator-prey relationship is also emphasized by introducing known predators into natural food environments. Food "chains" and food "web" are also "invented" by the teacher.

Teacher roles *B*, *C*, and *D* are again a part of the various activities outlined in the teacher's guide.

Relative Position and Motion

This fourth-grade unit is one of the most "teacher-structured" (more emphasis on Roles *B* and *C*) of all 12 SCIS units. Demonstrations and discussions are centered around relative positions of static systems and the motions of various moving systems. A pressed paper "stick" figure is used in various rotations and translations for relative position studies as well as in relative motion studies. The student manual is identified as "very important" to the activities in this unit and tends to add to the rather structured teaching-learning environment. "Flip books" are designed by the children to depict motion. These minicartoons are intended to help depict the relationship between relative positions and motion.

Tracks of rolling spheres, colliding spheres, and other activities are also used. Polar and rectangular coordinates are used in various types of mapping activities conducted both indoors and outdoors.

Communities

As with all of the SCIS units, this fifth-grade life science unit is projected for about four months of activities. The community concept of a food-energy relationship among populations of plants and animals is emphasized through activities with simple plants and animals. The interaction between producers (plants) and consumers (insects and animals) is studied and recorded in quantitative form on charts, graphs, and such. Frogs are at the top of the food web in this scheme. The role of decomposition in the community concept is studied briefly and a community game "rounds out" the unit.

The teacher's role in all of this varies among discussion leader (Role *B*), activity director (Role *C*), and activity facilitator (Role *D*).

Models: Electric and Magnetic Interactions

Activities in this sixth-grade unit are intended to help students build a "model" of electric and magnetic phenomena. Interaction-at-a-distance is observed and magnets and energy transfer are studied using electric circuits. The teacher role is similar to that in other units already described.

Summary of Teacher Roles and SCIS

The exploration, invention, and discovery phases that are described in each of the 12 guides for the teacher are somewhat misleading. Each word suggests largely "self-directed" activity (student-structure) rather than "other-directed" activity (teacher-structure), but the teacher's guides reveal a good deal of teacher-directed activity. Specifically, the invention and discovery phases of the lessons are usually initiated and directed by the teacher. In these, the teacher invents new ideas by way of a discussion leader and children are guided to discover further applications of the invented ideas.

Elementary Science Study and Teacher Roles

The final science curriculum project to be used as an example of varying teacher roles is the Elementary Science Study (ESS).[3] ESS

[3] Further information on ESS can be obtained from the publisher, McGraw-Hill.

varies considerably in format from SAPA and SCIS in that a curriculum for each grade level is not already predetermined. Fifty-six different units were developed by the ESS staff, but no overall structure for grades K–6 is required. A teacher may choose a dozen or so different, unrelated units as the curriculum for his or her class and no particular order for teaching the units is imposed. As with SAPA, SCIS and some textbook series, kits of manipulative materials are correlated with each unit (or at least most of them) and can be ordered from a commercial distributor. Table 5–2 shows the unit titles, a brief description of each, the content, and recommended grade levels.

TABLE 5–2. **Elementary Science Study (ESS) Units (from ESS Chart)**

Title	Description	Content	Grade
1. Light and Shadows	Children use many objects including their own bodies to study shadows.	Physical Science	K–3
2. Growing Seeds	Children plant seeds and observe their development.	Biological Science	K–3
3. Match and Measure	Measurement using various devices.	General	K–3
4. Mobiles	Children study balancing through mobile making.	Physical Science	K–4
5. Life of Beans and Peas	Life cycles are observed by replanting the "fruit" of a growing plant.	Biological Science	K–4
6. Primary Balancing	Simple pan balances are used.	General	K–4
7. Butterflies	Children observe various stages of development.	Biological Science	K–5
8. Pattern Blocks	Children use six geometrical shapes. (wooden pieces)	General	K–6
9. Geo Blocks	Blocks of various shapes are used for building.	General	K–6
10. Eggs and Tadpoles	Children observe the development of eggs to tadpoles to frogs.	Biological Science	K–6
11. Tangrams	Various geometric shapes (seven pieces) are fitted into puzzle outlines.	General	K–8
12. Musical Instrument Recipe Book	Construction of many musical instruments from common materials.	General	K–9
13. Attribute Games and Problems	Classification games are played with various materials.	General	K–9
14. Animals in the Classroom	Methods for keeping animals in the classroom.	Biological Science	K–9
15. Spinning Tables	Effects of circular motion are observed.	Physical Science	1 and 2

TABLE 5–2. (*continued*)

Title	Description	Content	Grade
16. Brine Shrimp	Brine shrimp eggs hatch.	Biological Science	1–4
17. Changes	Rotting and decay as well as changes in nonliving objects are observed.	Biological Science	1–4
18. Printing	Printing press is used to produce "children's views."	General	1–6
19. Mirror Cards	Mirrors are used with pattern cards.	Physical Science	1–6
20. Pond Water	Organisms in pond water are observed.	Biological Science	1–7
21. Daytime Astronomy	Records of the sky are kept.	Earth Science	1–8
22. Sand	Sand is poured, piled, and weighed.	Earth Science	2 and 3
23. Structures	Soda straws, pipe cleaners, and the like are used to build structures.	General	2–6
24. Sink or Float	Sinking or floating of various objects.	Physical Science	2–7
25. Clay Boats	Clay is used to make boats.	Physical Science	2–8
26. Drops, Streams, and Containers	Children observe various liquids by pouring and other activities.	Physical Science	3 and 4
27. Mystery Powders	Powders are mixed with various liquids to cause reactions.	Physical Science	3 and 4
28. Ice Cubes	Melting rates and temperature lowering of liquids.	Physical Science	3–5
29. Rocks and Charts	Common rocks are observed and classified.	Earth Science	3–6
30. Starting from Seeds	Similar to "Growing Seeds" but more advanced.	Biological Science	3–7
31. Where Is the Moon?	Changing positions of the moon are observed.	Earth Science	3–7
32. Colored Solutions	Colored solutions are mixed to produce new colors.	Physical Science	3–8
33. Mosquitoes	Mosquitoes are raised under varying conditions.	Biological Science	3–9
34. Whistles and Strings	Tubing, straws, and strings are used to produce sounds.	General	4 and 5
35. Bones	Real bones are used to reconstruct skeletons of small animals.	Biological Science	4–6
36. Small Things	Simple microscopes are used to view many "small things."	Biological Science	4–6

TABLE 5–2. (*continued*)

Title	Description	Content	Grade
37. Tracks	Children study animal tracks for clues.	Biological Science	4–6
38. Crayfish	Crayfish are observed and studied.	Biological Science	4–6
39. Budding Twigs	Twigs develop buds and leaves.	Biological Science.	4–6
40. Animal Activity	Various behaviors of small animals are observed.	Biological Science	4–6
41. Earthworms	Growth and behavior of earthworms.	Biological Science	4–6
42. Peas and Particles	Children estimate the amounts of seeds, marbles, and other objects in various containers.	General	4–6
43. Batteries and Bulbs	Circuits are made with batteries, bulbs, and wire.	Physical Science	4–6
44. Optics	Properties of light are studied.	Physical Science	4–6
45. Pendulums	Swinging objects are studied.	Physical Science	4–6
46. Microgardening	Molds are grown and studied.	Biological Science	4–7
47. Senior Balancing	Advanced forms of balancing.	Physical Science	4–8
48. Behavior of Mealworms	Experiments with mealworms are devised.	Biological Science	4–8
49. Stream Tables	Water streams illustrate erosion.	Earth Science	4–9
50. Water Flow	Water systems are constructed from tubing, bottles, clamps, and other parts.	Physical Science	5 and 6
51. Mapping	Maps of the school, classroom, and other locations are constructed.	General	5–7
52. Heating and Cooling	Candle flames heat various objects and rates of heating and cooling are observed.	Physical Science	5–7
53. Balloons and Gases	Gases are collected and weighed.	Physical Science	5–8
54. Gases and Airs	Interaction of air and common substances.	Physical Science	5–8
55. Batteries and Bulbs II	Battery-operated gadgets are constructed.	Physical Science	5–9
56. Kitchen Physics	Children observe the properties of common liquids.	Physical Science	6 and 7

The role of the teacher in the various activity units varies because of the wide variety of units and because different groups of people developed the units. The overall role that is emphasized in most teacher's guides is Role *D*, activity facilitator. Some units stress Role *C* whereas a few upper-level units stress Role *E*, co-investigator.

Contrary to SCIS, the developers of ESS felt that no particular science content should have precedence over any other. ESS identified *interest* as the main criterion in determining what should be a part of the curriculum and at which grade levels.

summary

Five distinctly different roles for the teacher who wants to help children learn science were identified:

A. Lecturer and question-answerer.

B. Discussion leader.

C. Activity director and evaluator.

D. Activity facilitator.

E. Co-investigator.

Roles *A* and *B* encourage verbal learning of definitions, rules, and the like, whereas Roles *C*, *D*, and *E* are more closely associated with non-verbal learning that grows out of a child's actions. Most textbooks suggest Roles *A* and *B* since the "content" to be learned consists primarily of what is contained between the covers of the books. The extent to which student investigations are an integral part of the content will, of course, determine how much teacher time will involve Roles *C* or *D* or *E*.

Four "conditions" were also identified as important criteria for determining whether or not certain teacher roles are logically compatible. Three of the conditions are derived from cognitive development work and one is derived from a definition of science.

Condition One

There are wide variations in developmental levels among children in most classrooms and, thus, equally wide variations in their conceptualizing abilities.

Condition Two

Conceptual learning is tied closely to developmental level and occurs primarily as a result of a child internalizing his or her own actions on objects.

Condition Three

It appears that direct, verbal instruction has little effect in facilitating advancement of developmental levels and, thus, the ability of children to conceptualize problems, beyond their current developmental abilities.

Condition Four

Science is an attempt to make more sense out of one's environment by manipulating portions of that environment at a difficulty level consistent with one's conceptual abilities.

Each of these four conditions was compared with each of the five teacher roles in terms of apparent compatibility, and a table summarizing all of the results was constructed.

Curriculum materials can, to a large degree, restrict the teacher role. When the materials for the children are mostly printed, the role of the teacher is "pushed" strongly toward Roles *A* and *B*. When the materials are mostly manipulative, the role of the teacher is directed more toward Roles *C* or *D* or *E*, with some emphasis on Role *B* still being very possible. Three well-known elementary science curriculum projects, SAPA, SCIS, and ESS, were described briefly and compared in

TABLE 5–3. Five Teacher Roles Compared to the Four Learning Conditions

	Learning Condition (Incompatible = *I* Compatible = *C*)				Type of Learning
Teacher Role	**1**	**2**	**3**	**4**	**(Verbal = *V*) (Nonverbal = *N*)**
A	*I*	*I*	*I*	*I*	*V*
B	*I-C*	*I*	*I*	*I*	*V*
C	*I-C*	*I-C*	*I-C*	*I-C*	*V* & *N*
D	*C*	*C*	*C*	*C*	Mostly *N*
E	*C*	*C*	*C*	*C*	*N*

terms of teacher role(s). Each curriculum is activity centered, but the emphasis on teacher role varies greatly:

1. SAPA emphasizes Role *C* with some of Role *B* in most lessons.
2. SCIS emphasizes Roles *C*, *B*, and *D*, respectively, at various points in the development of a unit of study.
3. ESS emphasizes Roles *D* and *C*, respectively, depending on the particular unit.

Very little use of Role *E*, co-investigator, is made by any of the science curricula in use in today's elementary schools, and, in fact, the major emphasis is with Role *A* or *B* or *C* or some combination thereof. We have seen that the four learning conditions are compatible with Roles *D* and *E* (activity facilitator and co-investigator) but incompatible or only partially compatible with Roles *A*, *B*, and *C* (lecturer and question–answerer, discussion leader, and activity director, and evaluator).

Appendix *A* contains a summary of two quantitatively defined teaching roles and their effects on various student behaviors and outcomes. After reading the description of the study, it should be fairly easy to see that Student-Structured Learning in Science (SSLS) corresponds roughly to Roles *D* or *E*, whereas Teacher-Structured Learning in Science (TSLS) corresponds to some combination of Roles *A*, *B*, and *C*.

What this boils down to is simply this: It is the teacher variable that really counts! Whatever role the teacher chooses to assume will largely determine the nature of the science program for children. The nature of children and the nature of science point toward Roles *D* and *E*. Other factors might point toward Roles *A* or *B* or *C* but the two most critical factors remain: children and science.

references

ALMY, MILLIE ET AL. *Logical Thinking in Second Grade.* New York: Teachers College Press, 1970.

BEILIN, HARRY, AND I. FRANKLIN. "Logical Operations in Area and Length Measurement, Age, and Training Effects." *Child Development,* **33**, 607–618, 1962.

BRAINERD, CHARLES, AND T. ALLEN. "Training and Generalization of Density of Conservation: Effects of Feedback and Consecutive Similar Stimuli." *Child Development,* **42**, 693–704, 1971.

EMERY, JOYCE. "The Status of Certain Probabilities Concepts and Combinatorial Abilities of High School Biology Students and the Effects of Genetics Instruction on these Cognitive Characteristics." Unpublished Ph.D. Dissertation, Department of Science Education, Florida State University, Tallahassee, 1973.

GOOD, RONALD ET AL. "Comparative Effects of Student-Structured Learning in Science (SSLS) and Teacher-Structured Learning in Science (TSLS) on Cognitive Development Levels of Elementary School Children." paper presented to The National Association for Research in Science Teaching, annual meeting in Chicago, 1974.

GRUEN, GERALD. "Experiences Affecting the Development of Number Conservation in Children." *Child Development,* **36,** 963–79, 1965.

HATANO, G., AND Y. SUGA. "Equilibration and External Reinforcement in the Acquisition of Number Conservation." *Japanese Psychological Research,* **11,** 17–31, 1969.

INHELDER, B., AND H. SINCLAIR. "Learning Cognitive Structures." In *Trends and Issues in Developmental Psychology,* edited by Mussen et al., pp. 2–21. New York: Holt, Rinehart and Winston, Inc., 1969.

LESKOW, S., AND C. SMOCK. "Developmental Changes in Problem-Solving Strategies: Permutation." *Developmental Psychology,* **2,** 412–22, 1970.

MURRAY, FRANK. "Cognitive Conflict and Reversibility Training in the Acquisition of Length Conservation." *Journal of Educational Psychology,* **59,** 82–87, 1968.

SMEDSLUND, JAN. "The Acquisition of Conservation of Substance and Weight in Children: II. External Reinforcement of Weight and of the Operations of Addition and Subtraction." *Scandinavian Journal of Psychology,* **2,** 71–84, 1961a.

———. "The Acquisition of Conservation of Substance and Weight in Children: III. Extinction of Conservation of Weight Acquired Normally and by Means of Empirical Controls on a Balance." *Scandinavian Journal of Psychology,* **2,** 85–87, 1961b.

———. "The Acquisition of Conservation of Substance and Weight in Children: V. Practice in Conflict Situations without External Reinforcement." *Scandinavian Journal of Psychology,* **2,** 156–60, 1961c.

———. "The Acquisition of Conservation of Substance and Weight in Children: VI. Practice on Continuous Versus Discontinuous Material in Conflict Situations without External Reinforcement." *Scandinavian Journal of Psychology,* **2,** 203–210, 1961d.

———. "Patterns of Experience and the Acquisition of Conservation of Length." *Scandinavian Journal of Psychology,* **4,** 257–64, 1963.

WALLACH, LISE, JACK WALL, AND LORNAL ANDERSON. "Number Conservation: The Roles of Reversibility, Addition–Subtraction, and Misleading Cues." *Child Development,* **38,** 425–42, 1967.

part six

CONCRETE IDEAS FOR HELPING CHILDREN LEARN SCIENCE

objectives

It is intended that this section will help the reader to

1. Identify a physical setting in the classroom that can facilitate the process of helping students learn science.
2. Identify specific classroom conditions and ideas for investigations necessary in helping students learn science through *growing plants.*
3. Identify specific classroom conditions and ideas for investigations necessary in helping students learn science through *raising insects.*
4. Identify specific classroom conditions and ideas for investigations necessary in helping students learn science through *small animals.*
5. Identify specific classroom conditions and ideas for investigations necessary in helping students learn science through *fishes, frogs,* and *snails.*
6. Identify specific classroom conditions and ideas for investigations necessary in helping students learn science through *liquids.*
7. Identify specific classroom conditions and ideas for investigations necessary in helping students learn science through *foods.*
8. Identify specific classroom conditions and ideas for investigations necessary in helping students learn science through *heating and cooling.*
9. Identify specific classroom conditions and ideas for investigations necessary in helping students learn science through *circuits and things.*
10. Identify specific classroom conditions and ideas for investigations necessary in helping students learn science through *closed systems.*
11. Identify specific classroom conditions and ideas for investigations necessary in helping students learn science through *making structures.*
12. Identify specific outdoor conditions and ideas for investigations necessary in helping students learn science through *bubbles, balloons, and paper planes.*

13. Identify specific outdoor conditions and ideas for investigations necessary in helping students learn science through *maps and measures.*
14. Identify specific outdoor conditions and ideas for investigations necessary in helping students learn science through *just watching.*
15. Identify specific outdoor conditions and ideas for investigations necessary in helping students learn science through *how many and how much.*
16. Help students learn science!

introduction

Part Two in this book presented a fairly detailed description of the general cognitive characteristics of children, from the time of early childhood through adolescence. It would be possible for the casual "peruser" of this book to assume, after reading Part Two, that science for children should consist of "diagnose what is missing in the student" and then "prescribe the remedy." If, for example, the ten-year-old is diagnosed as a nonconserver of weight, then that child should be given the necessary "prescription" to cure this ailment. This is just not the case.

A recurring theme in Piaget's theory of intellectual development negates the simplistic "diagnose-prescribe" cycle as a means of helping children develop logical thought. This theme could be called the interrelatedness or holistic nature of the development of intelligence. Cognitive development proceeds as an interrelated whole, with one piece or function linked to all others. One "piece" of the cognitive system cannot be developed (accelerated) at the expense of the remainder. All of the system develops as a total package. This means that a specific piece of the developing intellect must develop in conjunction with all other pieces, resulting in a cognitive system that is totally interrelated.

The example used earlier, of the ten-year-old who was a nonconserver of weight, can be restated again. It cannot be presumed that a "prescription" composed of weight experiences will cure this child's deficiency. It is the entire cognitive system that must advance, not merely an isolated set of mental structures that might be termed weight conservation. This explains why training studies have been largely unsuccessful in accelerating the cognitive development of children. Since the entire cognitive system must advance as an interrelated whole, it is illogical to try to accelerate a separate component (for example, classification, seriation, weight conservation, time).

It is this interrelated or holistic nature of intellectual development that negates the narrowly defined diagnostic teaching approach commonly used in reading, writing, and arithmetic. Part Two should have provided a fairly comprehensive picture of specific cognitive characteristics involved in the development of logical thinking that are used in the field of science. The total mental abilities of an individual are used in trying to figure out possible solutions to real problems in science. A curriculum should facilitate this interrelated approach in providing children with opportunities to use all facets of their mental abilities. One process or concept should not be emphasized to the exclusion of others. Problems that are real and interesting should be derived by the child from an environment rich in opportunities to probe, wonder, and experiment, in endless cycles, and at a level of difficulty that is consistent with current developmental abilities. The format of Concrete Ideas for Helping Children Learn Science is consistent with such an environment.

Review of Learning Conditions

Ideas for helping children learn science should grow out of a sound rationale. Goals should be included in the rationale and what becomes a part of the planned curriculum must be logically consistent with the goals. Thus far, we have considered the nature of children (primarily the cognitive developmental aspects), the nature of science (content-process), and the role of the teacher. Out of all of this, four conditions or considerations developed that give overall direction to an adult who wants to help children learn science. Briefly stated, they are as follows:

1. There are wide variations in developmental levels among children in most classrooms and, thus, equally wide variations in their conceptualizing abilities.
2. Conceptual learning is tied closely to developmental level and occurs primarily as a result of a child's internalizing his or her own actions on objects.
3. It appears that direct verbal instruction has little effect in facilitating advancements of developmental levels and, thus, the ability of children to conceptualize problems beyond their current developmental abilities.
4. Science is an attempt to make more sense out of one's environment by manipulating portions of that environment at a difficulty level that is consistent with one's conceptual abilities.

When five different teacher roles were compared with each of these four conditions, it became apparent that two of them were compatible with the four conditions, two were incompatible, and one was partially compatible. The two compatible teacher roles, which were identified as "activity facilitator" and "co-investigator," are emphasized throughout this chapter on concrete ideas for helping children learn science.

Grade Placement

Two important questions on "placement" of science investigations are:

1. Can children do it?
2. Will it make sense to them?

Particularly for the second question, it is important to understand the implications of the material in Chapter Two. A person's conceptual learning level is determined by that individual's cognitive development level. The range of developmental levels within any given grade level is 2 to 3 years. This means that conceptually appropriate science investigations must allow children to operate at their own levels. A child *cannot* be forced to *make sense* of that which is beyond his or her developmental level, nor *should* a child be forced to engage in learning activities that are not sufficiently challenging. The upper level of difficulty of most science investigations is open-ended. For example, an investigation dealing with electric (or electronic) circuitry can become exceedingly complex and abstract. The upper level of difficulty is open or not limited. A *lower* level of difficulty, however, is usually apparent with a given set of materials to be used in the investigation. If some flashlight batteries, bulbs, wire, switches, or the like comprise the set of materials, then a "beginning" level of difficulty is apparent if children try to light the bulb. Below the age of nine or ten years, most children exhibit very little systematic investigative behavior. A haphazard trial-and-error approach is used and unless a "correct" solution is stumbled upon by the child, the bulb will not be lighted. Even after this age, the development of systematic investigative behavior is exceedingly slow. Very few children at the elementary school level are capable of a systematic, quantitative approach to real problems and investigations.

For these reasons, many of the following ideas for investigations contain comments about the conceptual levels.

Classroom Preparation

Preparing a science program requires mental as well as physical preparation. This section contains a few suggestions for preparing for an *active* classroom where children are encouraged to *do science.*

1. If storage space is a problem, get some small cardboard boxes and stack them up along a wall as in Figure 6–1.

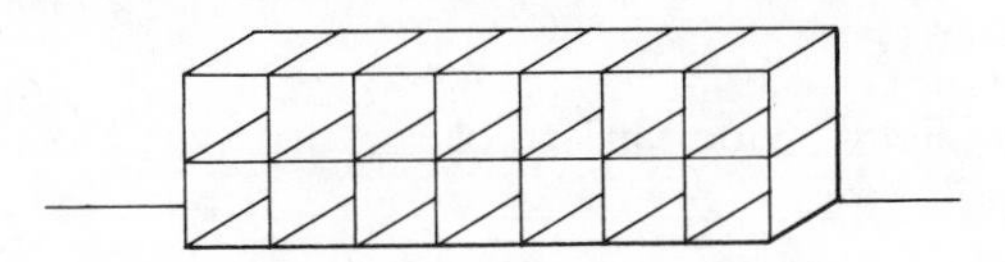

Figure 6–1. Classroom storage.

Most stores are very cooperative in saving "throw-aways" such as the cardboard storage bins. Tape the bins together and then to the wall. If the sizes of the bins are approximately the same, the storage unit should be fairly stable. A few smaller units around the room can also be arranged if space along one wall is limited. These cubicles can be used by the children for storing ongoing experiments that cannot be disassembled at the "end" of science time. The children can also store materials for future investigations in them.

2. Collect empty milk containers of various sizes. The one-gallon size is particularly useful for a wide variety of experiments. The containers can be used for growing plants, insect habitats, small animal habitats, aquaria, terraria, storage bins, and so on. Stringing the containers from the ceiling is a good "out of the way" method of storage. Various examples of the use of such containers will be evident throughout the examples of science investigations.
3. From a store that sells pianos and organs, try to get a couple of the heavy-duty cartons that were used for shipping. These cartons make great large storage bins or "special" places for experiments where light or sound can be blocked out. This is not a critical item but it can be useful and the construction of these cartons ensures a fairly long life in the classroom.
4. Many other items can be useful in implementing an *active* science program. Cities of even moderate size usually throw away more useful items in one day than can be used in a classroom during an entire year! A partial list of some of these "throw-aways" that can be very useful in helping children learn science is now given, with the corresponding locations.

Sources of Free Materials[1]

Supermarket—boxes and various containers, scrap meat and bones

Drugstore—eyedroppers, pill bottles, syringes.

Lumberyard—wood scraps.

[1] You are encouraged to extend the list of sources and add items to the existing sources.

Paint store—sample colors.

Paper or printing company—wide variety of scrap paper, old type, ink.

Restaurant—large jars, straws.

Tile store—odd pieces of floor and wall tile.

Newspaper plant—extra paper on rolls.

Junkyard—scrap metal, gauges, wire.

Computer location—punch cards, printout paper.

Hospital—pill bottles, syringes, plastic drinking cups, old x rays.

Rug store—samples and scraps.

Appliance store—old refrigerators, old motors, stoves or hot plates.

Seed store—assorted seeds.

School cafeteria—containers of various types.

Home—gallon plastic milk containers.

Office supply store—old electric adding machines, typewriters.

Leather store—scraps of leather.

Plastics store—plexiglas scraps, odd containers.

Nursery—plants, seeds, various soil mixtures.

Pond—tadpoles, fish, frogs, mosquitoes.

Outdoors—unlimited supplies.

indoor science

Even though most science programs could be conducted outdoors (weather permitting), many and varied reasons tend to keep even the active science programs indoors most of the time. Some investigations require that children go out-of-doors and many of these are considered in the following section on "outdoor science." We first look at a few ideas for indoor science.

Investigation Number One: Growing Plants

Growing plants is mostly an ongoing activity that will not require much time each day after the initial preparations are completed. Children in all of the elementary school grades can grow plants and enjoy

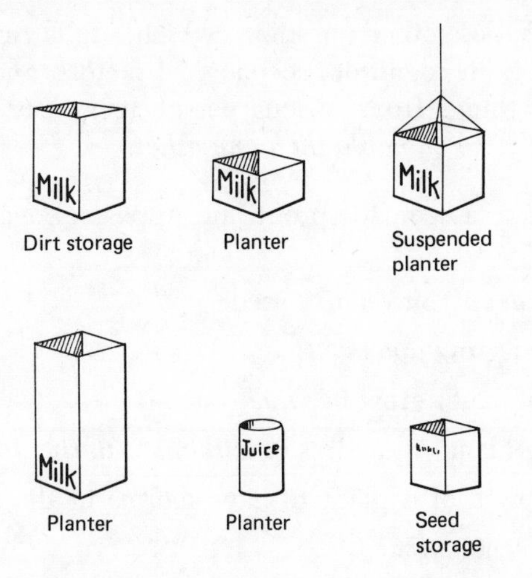

Figure 6–2. Materials for growing plants.

the process throughout the year. Having many plants around a classroom makes it much more "livable." Young children will not do much in the way of systematic, quantitative investigations, but they are quite capable of growing plants and gradually learning about factors that affect the growth of plants. Children will begin to show some signs of thinking about factors that affect plant growth by about third grade, although wide variations will exist among children's abilities to "investigate" plant growth.

Materials

Empty milk and juice containers, three or four different types of soil, seeds of various types.

Possible Activities

The activities listed here can vary considerably in the level of sophistication shown by children in "doing" the activity. The role of the teacher as *facilitator* is to encourage children to do what they can to find out what makes sense to them. All lesson-related behavior should be accepted. It is very important to encourage children to ask their *own* questions and try to find their *own* answers. Give suggestions *only* when you are *sure* a child is unable to think of activities and questions of his or her own. Teacher directions and evaluations encourage children to become dependent upon the teacher. Encourage independence by allowing children to follow their own ideas. Some of the questions that children may ask are listed. Countless other questions, which adults cannot think of because of differences in cognition, sociological factors, and an increasing rigidity in seeing things from various perspectves, may also be asked. *No one question is more valuable than any other.*

1. How *long* does it take a seed to sprout and appear above the ground?
2. Does it matter how *deep* you plant a seed?
3. Can you make a seed grow *down?*
4. Can a seed (and the plant) grow in *sand?*
5. Can a seed (and the plant) grow in soil *without watering?*
6. Does it help the growth of a plant to have *worms* in the soil?
7. Will a seed sprout in the *dark?*
8. Will a plant grow in the *dark?*
9. Will a seed in soil sprout in a *refrigerator?*
10. Will a plant grow in a *refrigerator?*
11. Will seeds that have been kept in a *refrigerator* sprout in soil?
12. What happens when *salt water* is used to "water" a plant?
13. Do some plants grow *faster* than others?
14. What happens if seeds are first *soaked* in water and then planted?
15. What happens if plants are watered with a *food coloring* solution?
16. Can a plant grow *upside down?*
17. Can you find plants that grow in *water?*
18. How do plants respond to *heat?*
19. How do plants respond to *sounds?*
20. Do plants *move* during the day?
21. Will a plant grow new leaves if some are *removed?*
22. What happens if part of the root system is *removed?*
23. What happens if an atomizer is used to spray *oil* onto a plant?
24. Does *smoke* affect the growth of plants?
25. Do all plants have some type of *flower?*
26. Can plant "*parts*" be used to grow new plants?

27. Will a plant grow in a "*Baggie?*"
28. How much *weight* do plants "gain" each week?
29. How much *growth* occurs each week?
30. Can plants be grown using light bulbs of various *colors* as the only light source?
31. Which grows *fastest:* carrots, beans, radishes, tomatoes, potatoes, corn, rice, wheat, or turnips?
32. Could a carrot be "*crossed*" with a radish to get a "carrod" (radcar)?
33. Do the *insides* of seeds look alike?
34. Will deodorants and hair sprays affect plant growth?
35. Can *half* a seed sprout?

These are only a few of the many questions about plants that children of elementary school age may ask. Typically, younger children will not identify questions in advance of a considerable amount of *doing*. Questions will likely develop as a result of the child's noticing certain interesting things that grow out of his or her relatively random, unplanned activity. Careful investigations will definitely not occur, but some "experimentation" will be noticed. Even the ten and eleven-year-old will be haphazard in trying to find answers to questions, but the older child is much more likely to *ask* some questions in advance of random activity with materials. Older children are also more likely to respond to teacher questions by making some attempt to suggest a possible way of finding out. Measurements will not be very exact and many variables will probably not be controlled, but experiments can be designed in an attempt to "find out."

Investigation Number Two: Raising Insects

Raising insects can be a very high interest activity because of the active nature of many insects. The term *insects* is used rather loosely, and includes all "crawly critters" regardless of body sections, legs, and wings (or lack thereof). Insects are widely available and relatively easy to maintain in the classroom. They are more readily available in the fall and spring months, but they can be located even in the cold winter months. Keeping insects alive in the classroom merely requires small habitats in which the interiors are similar to the insect's natural environment. Needed materials and sketches of some possible habitats for insects are now described.

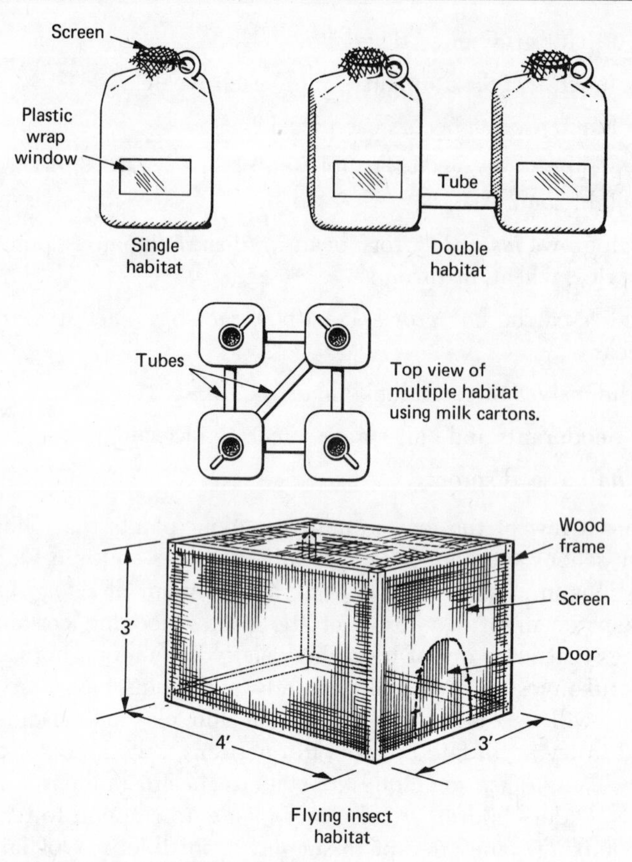

Figure 6–3. Habitats for insects.

Materials

Empty milk and juice cartons of various sizes, plastic wrap, tape, toilet paper rolls, soil, plants, apples, and potatoes (for moisture in habitats).

Possible Activities

Since a child's conceptual abilities will largely determine his or her approach to finding out, specific investigative suggestions are again avoided in favor of questions. A question can be pursued at many levels of sophistication, depending on what makes sense to a given child. It is important to keep in mind that the following questions are merely *possibilities* that children might make up. Each child should be encouraged to make up his or her own question (s) and then try to figure out a way of finding out. Most children will tend to do a lot of "messing around" before any questions arise. This is a *natural* and *necessary* process. If the children's activity reveals *high interest,* it indicates productive behavior on their part. Remember, *no one question is any more or less important than any other.* Most questions are not restricted to a specific insect, although sometimes certain insects are referred to.

1. Can small insects such as ants survive in an *airtight* terrarium?
2. What *foods* do insects seem to prefer?
3. Do *plants* seem to help insects survive in their habitat?
4. What happens if the habitat is kept in the *dark* for a few days?
5. How long can insects survive in a *cold* place (refrigerator, ice chest)?
6. Do insects prefer one kind of *soil* over another?
7. What seems to be the *life span* of various insects?
8. Do insects seem to give indications that they are *asleep?*
9. What do larger insects look like *inside* (after they die)?
10. Can you see insects *breathe?*
11. How many *legs* do most insects have?
12. Can an insect *back up?*
13. Do insects give any indications that they can *hear?*
14. What is the *fastest* crawling insect that you can find?
15. Do insects seem to prefer their food *wet* or *dry?*
16. Do some insects live in *water?*
17. Do all *flies* look alike?
18. Do most insects lay *eggs* to produce new insects?
19. How long do the eggs of various insects take to *hatch?*
20. Does an *earthworm* have eyes?
21. Do some insects not have *eyes?*
22. What kind of *soil* do worms prefer?
23. Do insects seem to prefer some *colors* over others?
24. Can big worms *crawl* faster than small worms?
25. Do mealworms (get them at a pet store) seem to prefer some *cereals* over others?
26. Can different insects live together in the same *habitat?*

27. Do larger insects seem to *live longer* than smaller ones?
28. What do worms *eat*?
29. What happens if a *spider* is introduced into an insect habitat?
30. What happens if other insects are introduced into a *spider* habitat?
31. Can some insects *smell* foods?
32. Can a mosquito fly *backward*?
33. Does *honey* really attract insects?
34. Will a spider *repair* its web if a hole is torn in it?

Although constructing the habitats for the insects can be a bit difficult for younger children, it is as valuable for them to do this as experimenting with the insects. Helping the children build the habitat should thus be avoided unless it seems absolutely necessary. Cutting the "windows" out of the plastic or cardboard containers may be a step best done by the teacher to avoid serious cuts. As a facilitator, decisions are constantly made by the teacher about how to facilitate.

Most children, and in particular the younger ones, do not really "experiment" in any careful, quantitative sense. *There are never experiments that fail!* Sometimes expectations are not realized, but that is certainly not a failure. When children find different answers to a question, your role as the teacher (facilitator) is not to decide who is most correct. The method of "finding out" should be emphasized. *All of the children should be encouraged to consider each other's methods of finding out, rather than worrying about correct answers.* The facilitator encourages openness and further inquiry rather than closure. There are never final answers in science. Ideas are continually revised as more data are collected. It is an open-ended process.

Investigation Number Three: Small animals in the classroom

The mere fascination of having small animals in the classroom is worth the effort needed to prepare habitats for them. As with the first two investigations, it is a good idea to begin this activity as early as possible and continue throughout the school year. This makes it possible for the children to observe life cycles in animals and, perhaps, to experiment with crossing different characteristics (for example, black mice with white mice).

Studying the "social" characteristics of animals under various conditions also takes some time. Constructing the habitats can be done both at home and at school. Old aquaria make excellent habitats as do interconnected wood and screen containers.

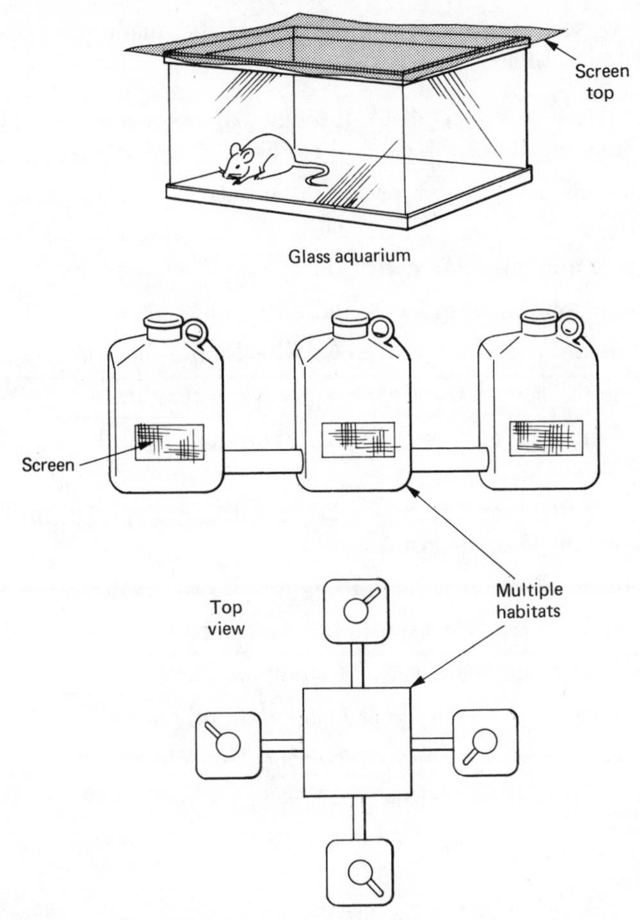

Figure 6–4. Small-animal habitats.

Materials

Aquaria, wood and nails, screen or wire mesh, masking or carpet tape, water bottles, wood shavings, heavy plastic or cardboard tubes, food pellets, exercise wheels. Some examples of habitats for small animals such as mice, gerbils, rats, hamsters, and snakes are shown in the following sketches.

Possible Activities

Since the construction of the habitats is a very important part of the overall set of activities, some of the questions refer to habitat construction. Younger children are less capable of *constructing* habitats from wood and screen so every effort should be made to provide easily constructed habitats.

1. If habitats are *connected* by tubing, wire screen, and the like, will the animals move back and forth?
2. Do animals seem to *prefer* glass habitats (aquaria) or screen habitats?
3. Are some animals more *active* than others?
4. Do some animals seem to *eat* more than others?
5. Do some animals use the *exercise* wheel more than others?
6. Do animals seem to use their *tongue* when they drink?
7. Is the animal's *tail* used for anything or does it just "drag along?"
8. Do the animals seem to prefer being out in the *open* or inside a small box or other enclosure?
9. Do some small animals seem to *fight* more than others?
10. How often do animals seem to need a *drink*?
11. Can you find the *heartbeat* of a small animal?
12. Do some animals' *hearts* beat faster than others?
13. Do some animals have more *offspring* than others?
14. Do both the mother and father *care* for the offspring?
15. How long does it take for a *newborn* animal to become full grown?
16. How *heavy* is a newborn animal?
17. If a newborn animal *dies,* could you find out why?
18. Will small animals *chew* up wood and plastic?
19. Do the animals seem to prefer *food* in the form of pellets or as a powder?
20. Can some animals get through a *maze* to some food faster than others?
21. If an animal goes through a *maze* a number of times, will it succeed in doing so in less time?
22. Do animals seem to prefer some *colors* over others?
23. If black mice are always *crossed* with white mice, what do the offspring look like?
24. If an adult male animal and an adult female animal are placed

together in a large cage (at least $2' \times 2' \times 2'$), how many offspring will they produce in 3 months?

25. If, in Question 24, the *offspring* are kept in the cage and allowed to mate, how many animals will there be after 6 months?
26. How much *weight* does a baby animal gain each day?
27. Will animals still move about if their habitat is covered with a *dark* cloth?
28. Will animals eat their *food* hot or cold?
29. Will animals *eat* the same kind of food as people eat?
30. Can an animal walk a *tightrope?*
31. What will an animal do to a plastic drinking *straw?*
32. What happens to an empty cardboard *juice can* (concentrate) when it is placed in the animal's cage?
33. Will animals prefer such *drinks* as Kool-aid, orange juice, lemonade, coffee, tea, or soda pop to water?
34. Do animals have different *footprints?*
35. Does the length of a mouse's *tail* have any relationship with its body weight?

For every question or observation that adults can think of, children will "stumble across" many more. Their questions usually grow out of their actions so it is wise to encourage lots of *actions first.* The level of the children's questions will be limited by their developmental level. Cause-and-effect questions will be very vague, inconsistent, and haphazard for young children and gradually become more specific and systematic as they develop the cognitive structures necessary for more logical thinking. When you as the teacher ask children questions, give them time to mull the ideas around for a few minutes. Explanations that are *made up* in the mind of the child usually take some time to formulate, whereas *recalling* explanations is a much faster process.

Investigation Number Four: Fishes, Frogs and Snails

If the previous three investigations plus this one are begun during the first few weeks of school, children will have an opportunity to observe and experiment with the organisms throughout the year. Where growth is slow or life cycles are involved, time is needed for noticeable changes to occur.

Living systems are among the most complicated systems to study in any carefully controlled, quantitative sense. They provide great opportunities, however, for children to ask questions and then figure out ways of finding out. The primary purpose of the activities is not to learn facts about flies or fleas or fishes or frogs or snails, although that will inevitably occur, but is for the child to learn how to find out what he or she wants to know by getting in there and doing it. Self-confidence for independent learning grows as the individual experiences the success and joy of finding things out. To facilitate that process, you must be sensitive to each child and encourage further exploration in all investigations. There are *no final answers,* only temporary solutions that must be revised constantly as the individual grows *and* as more is learned about the system of study.

Materials

Aquaria of various sizes, magnifying glasses, gallon glass jugs (wide mouth), fishes, frogs or tadpoles, snails, filter and pump if possible, water plants, small fish net, glass containers of various sizes for holding individual fish, frogs, and others.

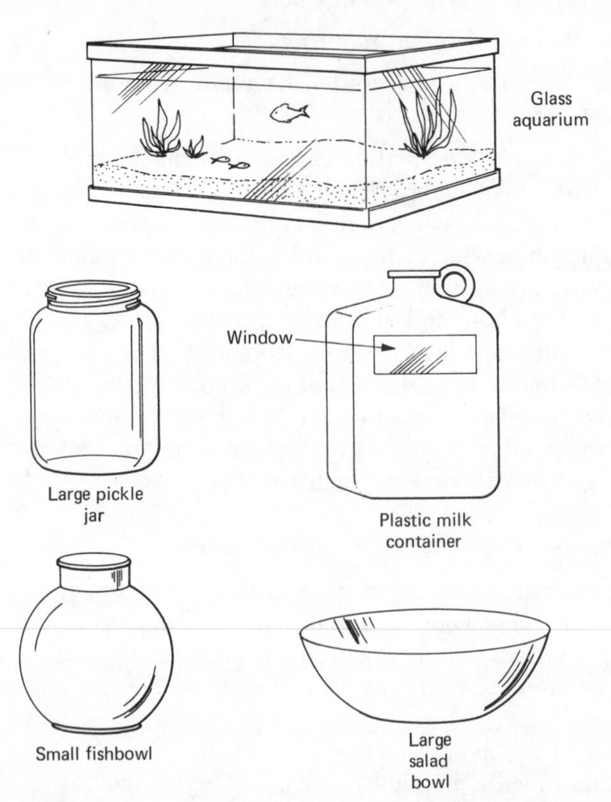

Figure 6–5. Materials for fishes, frogs, and snails.

Possible Activities

These aquarium creatures provide opportunities for observing and studying the development and behavior of aquatic life throughout the year. The frogs and snails in particular will produce large numbers of developing eggs that can be studied by every child. Having all of the animals in the same tank, along with aquatic plants, can show the "natural" interaction among living things. If microscopes are available, other water organisms might be discovered in addition to the more obvious fishes, frogs (tadpoles), and snails. If possible, local pond water should be used for at least some of the aquaria. This water will contain some of the tiny creatures that are killed in the purification process at water plants. Tap water can be used for aquaria, but caution should be used when small organisms (such as snail eggs) are involved.

1. Do *all* fertilized *frog* eggs develop into tadpoles or only *some* of them?
2. How many days elapse until the *tadpoles* develop *legs?*
3. If fish are in the same aquarium as the frog eggs, do the fish *eat* the eggs?
4. How long does it take for *snail* eggs to *develop* into moving snails?
5. Where do *snails* seem to spend most of their time?
6. Can you see any evidence that snails *sleep?*
7. Can you see any evidence that *fish sleep? Tadpoles?*
8. If there are *guppies* in the tank, how many *young* are usually born at one time?
9. How long does it take the *guppies* (or other "live-bearers") to become *full grown?*
10. Are *plants* in the aquarium eaten for food?
11. Do water plants have *roots?*
12. Do *snails* discard their *shells* as they grow larger?
13. Can *snails* live in *salt water?*
14. Can the freshwater *plants* live in *salt water?*
15. Can saltwater *plants* live in *fresh water?*
16. What happens if a *crayfish* is introduced into the aquarium?
17. If a crayfish loses a *pincer,* will the pincer grow back?
18. Do guppies and other small *fish* have *tongues?*

19. Can you think of a way to find out if a *fish* has a sense of *smell*?
20. Will fish eat a *worm*?
21. Do fish have *ears*?
22. Can you think of a way to find out if fish can see *colors*?
23. If a *turtle* is introduced into the aquarium, how do the other inhabitants react?
24. Do certain *fish* seem to stay in certain *places* in the aquarium?
25. If a few drops of *alcohol* are added to a *snail* aquarium, are there any noticeable effects?
26. If a few drops of liquid *soap* are added to a *snail* aquarium, are there any noticeable effects.
27. What happens as the *temperature* in a *snail* aquarium is lowered?
28. How long can a *snail* survive on an *ice cube*?
29. Do *snails* have *blood*?
30. Does a tadpole have a *heart*?
31. Can you find a *heartbeat* in a frog?
32. Will *frogs* eat *insects*?
33. Do *snails* have *eyes*?
34. Can a *snail* survive without its *shell*?
35. What happens if *food coloring* is added to an aquarium?

Investigation Number Five: About Liquids

The first four investigations dealt with living systems. Nonliving systems offer equally interesting opportunities for asking questions and finding answers. In many ways, nonliving systems can be less complicated to study because they do not act of their own volition. This means that a given nonliving system should behave in a highly predictable way, time after time *if* the same external factors are present. Children have a natural affinity for liquids so the motivation factor is high for this activity.

Materials

Water, cooking oil, vinegar, salt, sugar, liquid detergent, soap (bar), glycerin (if possible), ammonia (optional for young children), dishpan-size containers, buckets, small containers of various sizes,

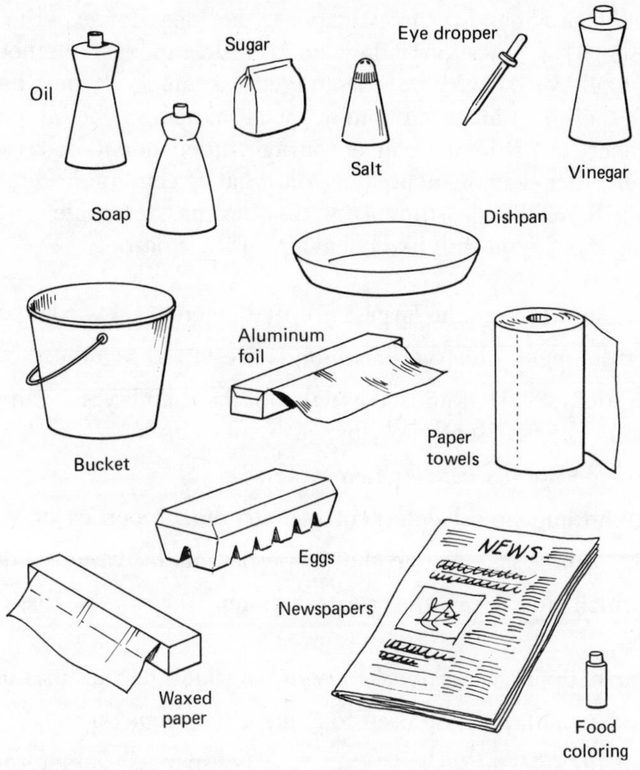

Figure 6–6. Materials for investigations about liquids.

flexible wire, wax paper, aluminum foil, lots of old newspapers, a few sponges, paper towels, eyedroppers, Ziplock bags, food coloring, egg cartons (styrofoam), various types of soda pop.

Possible Activities

The materials list could contain many more items and probably will as children begin to experiment with liquids. Young children will do a lot of mixing and cooking and "making believe" with a lesser amount of "trying to find out" behavior. This is quite natural and will usually lead to what appears to be more "productive" behavior in attempts to find out something about liquids. If there are limits to the type of messing around behavior, these should be established early and followed consistently. If the limits are fair and involve mostly disruptive behavior, most children understand why the limits are there and accept them. It is important that all children clean up any messes (have a mop handy) and learn to take responsibility for their actions.

As with the previous five investigations, the ideas for experimentation are presented as questions allow for a wide range of methods of finding out. Children should be encouraged, as much as possible, to come up with their own ideas and find out what happens. The interaction between teacher and students can encourage independent behavior or teacher-dependent behavior, depending on what is communicated. It is quite possible to challenge students with questions, example systems, and the like without *requiring* like behavior on their part.

1. What liquid forms the largest *drop* (diameter) on waxed paper?
2. What happens when various liquids are *mixed* together?
3. How do various liquids *compare* in their ability to "support" various objects (toothpick, pencil, paper clip)?
4. Are some liquids *heavier* than others?
5. Does adding liquid detergent change the *properties* of water?
6. Does adding *salt* change the properties of the various liquids?
7. In which liquid can the greatest amount of *salt* be dissolved? *Sugar?*
8. What happens when *food coloring* is added to various liquids?
9. Could some liquids be used to make a *thermometer?*
10. What liquid is *absorbed* fastest by newspapers? Paper towels?
11. Can all liquids be *frozen?*
12. Do *cold* liquids behave differently than *warm* liquids?
13. Does *heating* or *cooling* affect the dissolving ability of liquids?
14. How many *teaspoons* of water equal one quart? A gallon? A cup?
15. What *volume* of water equals the weight of the heaviest student in class? The lightest?
16. Which liquid "runs" *slowest* down a slope? *Fastest?*
17. Does the addition of *food coloring* affect the way water freezes?
18. Do some liquids *evaporate* faster than others?
19. Could bread be made with *salt water?*
20. What is the least amount of *sugar* that can be added to water so that it can be "tasted"? *Salt?*
21. Will a quart of water *freeze* faster than a gallon of water?
22. What happens when *corn starch* is mixed with water?

23. How much water *content* is in milk, grapefruit, apples, banana, tomato, and other foods?
24. Does Alka Seltzer *fizz* the same in salt water? Milk? Soda pop? Vinegar?
25. Does the *fizz* last just as long in various types of soda pop?
26. Does *temperature* affect the fizz time for soda pop?
27. What happens to soda pop if *salt* is added? *Sugar? Baking soda?*
28. How does the *freezing rate* of fresh soda pop compare with that of "flat" soda pop?
29. Is it possible to identify various types of *colas* merely by tasting them?
30. Is it possible to put the *fizz* back into soda pop?
31. Will some *metals* dissolve at different rates in various liquids?
32. How does *temperature* affect dissolving rate?
33. Do lighter liquids *freeze* faster than heavier liquids?
34. How does the addition of salt affect the *boiling point* of water? Sugar? Baking soda? Flour? Chocolate?
35. Is it possible to "*compress*" a liquid?

Investigation Number Six: About Foods

All of us know lots of things about the foods we eat. We know how they taste, smell, and look. A great many things, however, can be "found out" about foods that are typically not known. More important than a knowledge of specifics is the great opportunity to learn about the process of "finding out." A wide variety of foods is available to students and all that is needed is a different orientation toward finding out about foods. Albert Einstein once remarked that science is merely a refinement of everyday thinking. We can "refine" our everyday thinking about foods by deciding to do so. This particular area of investigation offers many opportunities for work at home as well as at school. It can be one of the most worthwhile, interesting investigations that can be pursued in science for children.

Materials

All kinds of foodstuffs—fruits, vegetables, meats, canned foods, breads, ketchup, mustard, mayonnaise, sugar, flour, and so on. Also, aluminum foil, Ziplock bags, petri dishes (if available), hot plates,

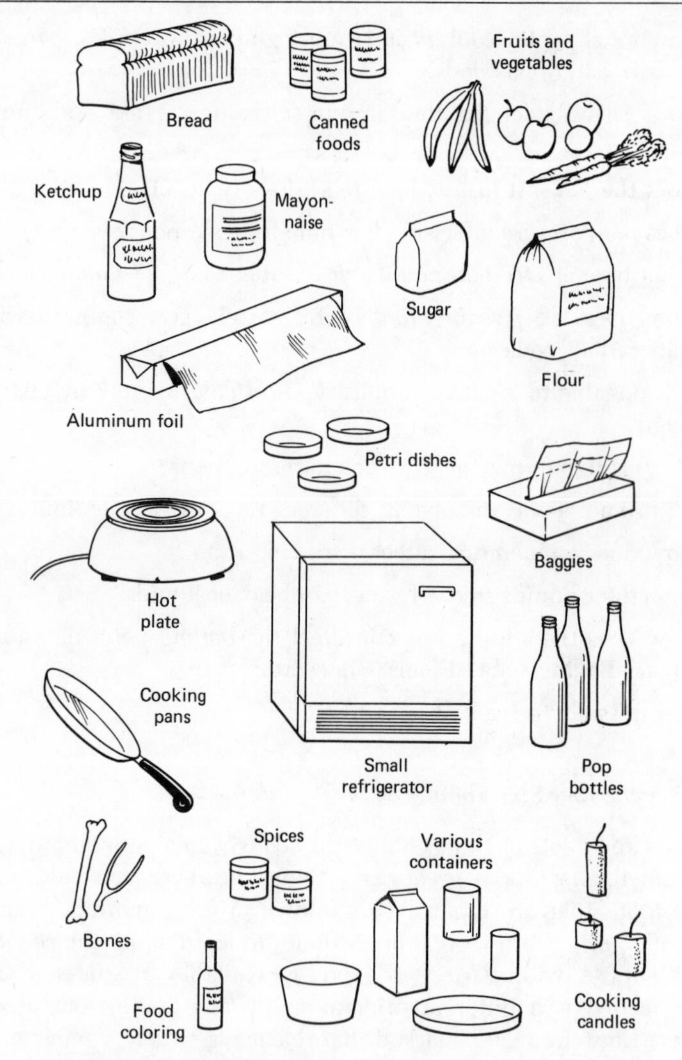

Figure 6–7. Materials for foods activity.

refrigerator (if possible), various sizes of mixing containers, plastic wrap, rubber bands, some pans for cooking or heating, empty soda pop bottles with plastic tops, food coloring.

Possible Activities

1. What is the "best" way to grow *mold* on bread?
2. Is it easier to grow *molds* on some types of bread than on others?

3. Will *mold* grow on meat?
4. Will *mold* grow on fruit?
5. Do all *molds* look the same?
6. What happens if *bread* is baked without sugar? Salt? Yeast?
7. What happens to *bread* if it is made with twice the amount of sugar in the recipe? Salt? Yeast?
8. What happens if the amount of *shortening* is varied in recipes?
9. Can *Jello* be frozen and thawed out?
10. Could chocolate *butter* be made?
11. If *food coloring* is used to color various foods, does it also affect their taste?
12. If people cannot see what they eat, does it affect their *taste?*
13. What ways can be used to prepare *dried foods,* such as those used for camping?
14. How long can *milk* be kept before it sours?
15. Does *freezing* affect the taste of some foods more than others?
16. If people cannot *smell* what they eat, does it affect their taste?
17. Are there ways to *identify* foods if they have been burned and charred?
18. Does a charred apple *taste* different than a charred pear?
19. Is the *taste* of foods affected by soaking the foods in water? Salt water? Sugar water? Pepper water?
20. What happens when various foods are *soaked* in water for long periods of time?
21. What happens if *sugar* is added to various kinds of fruit and the whole thing is kept in a closed container for a few days? Weeks?
22. What food is edible that grows in a *forest?*
23. What would happen if you changed to a *vegetarian* diet for a few days or weeks?
24. If a group of people *did not eat* for a whole day, would they all lose the same amount of weight?
25. Could you concoct a *recipe* for making *soup* from fresh vegetables?
26. Does meat *tenderizer* make all meats more tender?
27. What happens if a recipe for homemade *ice cream* is changed?

28. What do various foods look like under a *microscope?*
29. What are the best conditions for *growing molds* on food?
30. How much do watery *liquids* (milk, juices) *expand* when frozen?
31. Can regular *corn* be "popped?"
32. What is the best way to "*pickle*" a cucumber? A beet? Pig's feet? An apple?
33. If everyone in a classroom brought one kind of food for a *picnic,* without discussing who would bring what, what would be the result?
34. Which foods are easiest to *dye* with food coloring?
35. If you could have *three foods* to live on for the rest of your life, what would you choose?

Investigation Number Seven: Heating and Cooling

The properties of most substances change as their temperature is raised or lowered. Physical and chemical changes can be observed in many common substances by heating and/or cooling them. It is very interesting to just "see what happens" when something is placed in a freezer or heated with a candle or hot water bath. Some substances, especially certain liquids, can be dangerous if heated so it is important to warn students to avoid heating such liquids unless you, as the teacher, have checked the liquids out. Older students can probably take the proper precautions if they are made aware of potential harmful effects, but younger children are generally unable to look at various possibilities and take similar precautions. Using common sense with candles is really all that is usually necessary to avoid unfortunate accidents.

Materials

Food-warming candles or other "short and fat" candles, hot plate, glass beaker, ice chest with ice, thermometers, aluminum foil, plastic Ziplock bags, containers of various sizes, squeeze-type clothespins, foods of various types, kitchen powders, liquids, other "heatables" and "coolables."

Possible Activities

1. What happens to *sugar* when it is heated with a candle? Salt? Pepper? Baking soda? Flour?
2. What happens to *macaroni* when it is heated?

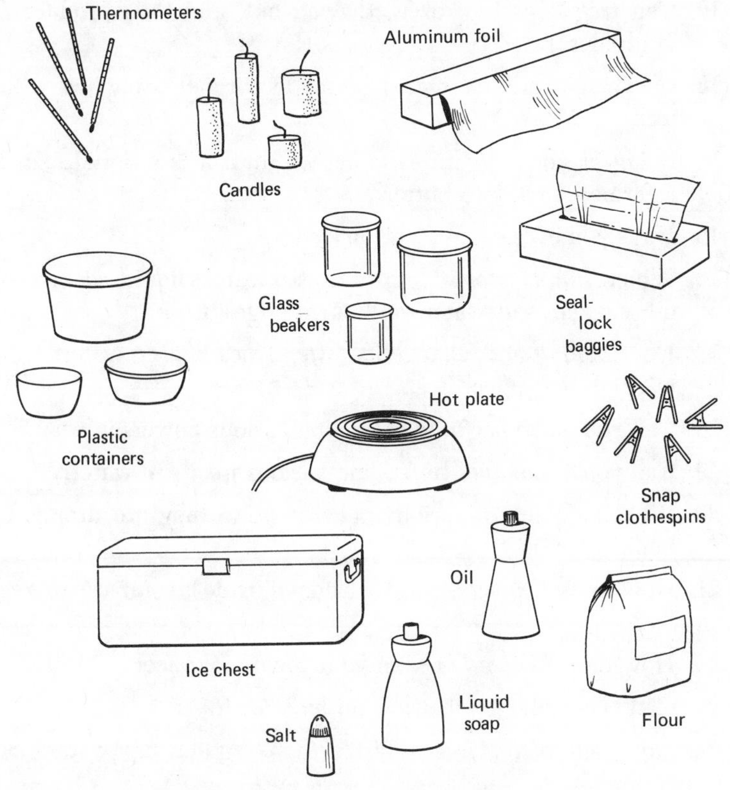

Figure 6–8. Materials for heating and cooling.

3. Do different kinds of *cloth* (very small pieces) burn the same way?
4. Can *honey* be frozen? Vinegar? Salt water? Sugar water? Liquid soap?
5. Does the *shape* of an ice cube affect its rate of melting?
6. Will an *ice cube melt* faster in sawdust or sand? Dirt? Salt?
7. If small *candles* are *melted* to make a big candle, is there any loss in weight?
8. What happens if hot wax from a *candle* is melted onto an *ice cube?*
9. Is *popped corn* just as *heavy* after it pops?
10. Do some metal *wires* "heat up" faster then others?
11. Does hot water *freeze* faster than cold water?

12. Can *frog eggs* be frozen, thawed out, and then produce frogs (tadpoles)?
13. Can homemade *ice cream* be made without using *salt* with the ice?
14. If *butter* (or margarine) is *heated* until it is a liquid, can it be reformed back into butter?
15. Have you ever eaten a hot *pickle?*
16. Which *liquid cools off* the fastest: water, liquid soap, cooking oil, vinegar, salt water, milk, or orange juice?
17. Does *boiling milk* change its *taste?* Does boiling water? Tomato juice? Prune juice?
18. Can you make *perfume* by boiling various flowers in water?
19. Can you make *dyes* by boiling various plants in water?
20. What happens to various *plastics* when they are dropped into boiling water?
21. How does *dry ice* (solid carbon dioxide) compare to regular ice?
22. How does the *ice house* in your town make ice?
23. Can you make *ice* that has no *bubbles* in it?
24. Does *salt water freeze* faster than regular tap water? Soapy water? Sugar water?
25. Could you use a *light bulb* to boil water?
26. As you get farther and farther from a *heat source* (light bulb, candle) how does the temperature vary?
27. As you get farther and farther from a *"cold" source,* how does the temperature vary?
28. How does water temperature affect the *melting rate* of ice in the water?
29. If a *pen* is *frozen,* does it still write?
30. What happens to a blown-up *balloon* if it is cooled? Heated?
31. Can you make a *hot-air balloon* that really works?
32. Does heating a *magnet* affect its "strength?"
33. Can *bone* be *burned?*
34. Can *blood* be frozen?
35. Can a chicken *egg* be frozen?

Investigation Number Eight: Circuits and Things

Making simple circuits with batteries, wire, bulbs, and things can be a truly challenging and fascinating activity. Many of the ideas for investigation, however, are probably a bit difficult for younger children (K–2) because of their tendency for random, trial-and-error efforts in problem solving. Discovering the combination of batteries, bulbs, and wires that "work" out of those that are possible requires a problem-solving strategy that is more advanced than a simple trial-and-error approach. There will always be variations of conceptual ability within any classroom and, thus, variations in ability to solve problems, even for older children.

Some children will probably go far beyond the simple circuits at the beginning. Old radios, motors, flashlights, and other electric devices will appear in the classroom and, of course, such interest should be encouraged. What begins as mild curiosity can grow into months of intense interest into the workings of telephones, motors, radios, record players, and countless other devices. Simple motors, radios, transmitters, buzzers, and similar electrical devices can be constructed with the help of guides to be found in the library or at specialty stores. Caution should be used with *any* device that uses a 110-volt power source rather than a battery. In fact, it is probably a good idea at the outset of this investigation to warn against plugging in any device that is new to a child unless an adult is present. It is one thing to plug in commonly used appliances at home but quite another to experiment with "junkyard" equipment that has not been tested.

Materials

Various types of batteries, flashlight and other low-voltage bulbs, lots of wire including some plastic coated "magnet" wire, rubber bands, aluminum foil, tape, cardboard, snap-type clothespins, small motors, magnets, buzzers and bells, old telephones, tin cans, old radios and small appliances, paper fasteners and clips.

Possible Activities

1. Can you light a *bulb* using only one battery and one piece of wire? Two bulbs?
2. How many *bulbs* can be lighted using only one battery?

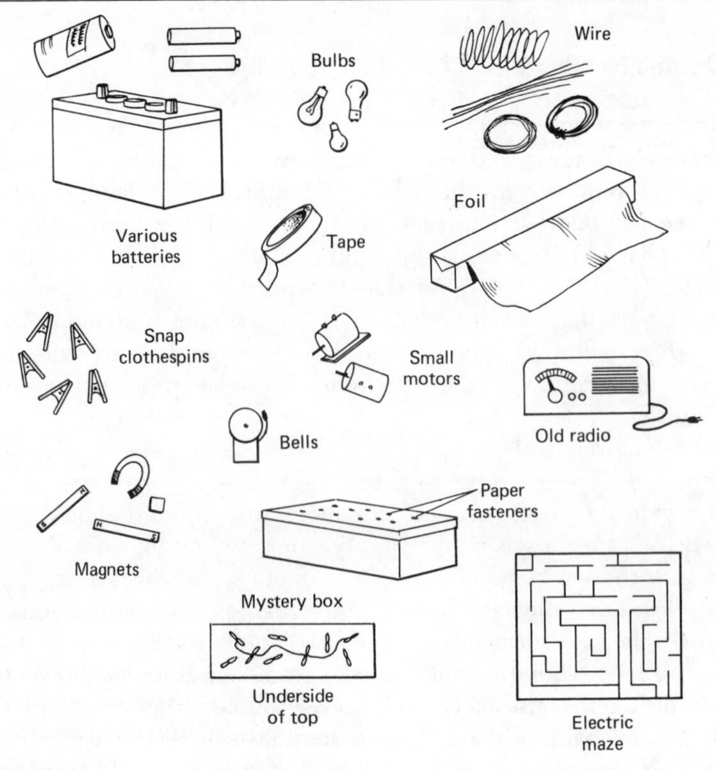

Figure 6–9. Materials for circuits and things.

3. Can *aluminum foil* be used to replace the wire in a simple circuit?
4. How many *different ways* can a bulb be lighted using one battery and two wires?
5. Are the *insides* of all *bulbs* the same?
6. Will a bulb work after the *glass* has been *cracked* or broken?
7. How many batteries are needed to *"blow out"* a bulb?
8. If one battery lights a certain number of bulbs, how many batteries are needed to light *twice* that number?
9. Can you make *drawings* of *circuits* that can be used by others to construct the same arrangements?
10. Can you use an old shoe box to make a *mystery circuit*? (see sketch following materials list).
11. Have you ever looked at the *insides* of a *battery*?

12. What happens if a *tin can* is inserted into a circuit? A key? Aluminum foil? Scissors? A paper clip?
13. Does the *length* of the *wire* in a circuit affect the brightness of the bulb(s)?
14. Will a circuit work *under water?* Salt water? Sugar water?
15. Can you *make* a *bulb* that really lights up?
16. Is a magnetic *compass* affected by a circuit?
17. Can you make an *electromagnet* using a large nail and magnet wire?
18. Does the number of *coils* in an electromagnet affect the strength of the magnet?
19. Can you make a *buzzer* using an electromagnet and a strip of metal? (see sketch)
20. Can you make a *bell* from a *telephone* work?
21. How *long* will one battery ring the bell? Two batteries? Three?
22. Can you figure out how to make a *motor* by looking at some small commercially available models?
23. Can you make a *switch* to turn a circuit on or off?
24. Can you make your own *flashlight?*
25. Can you make a *telegraph* and use it to send messages to other parts of the school?
26. Can you make an *electric maze* using aluminum foil? Other games?
27. How difficult would it be to make a *model* of your *school,* complete with lights, bells, and switches?
28. Is it possible to *test old radios* for "burned out" tubes or other defects using simple circuits with bulbs?
29. How many *different ways* can three batteries and three bulbs be connected so that all the bulbs are lighted?
30. Who is the best "*mystery circuit*" maker in your class?
31. Can you get *old gauges* from a junkyard to work in your circuits?
32. Can you find a way to make a simple *radio?*
33. How do the insides of a *radio* compare with those of a *television?*
34. Can you build a *static maker* that causes radio interference?
35. Can you visit some places in town that *repair* electronic devices?

Investigation Number Nine: Closed Systems

As with the batteries and bulbs investigation, this activity is suggested mainly for children who are at least nine to ten years old. The physical manipulations and especially the conceptual requirements of the systems to be studied pretty much rule out younger children. Closed systems investigations offer opportunities to make up explanations for certain variations, and some facility for causal thinking is needed. For this activity it is important that you as the teacher be able to construct various closed systems and have some understanding of their operation so that you can interact properly with your students. The sketches following the materials section should help somewhat in communicating the physical nature of various closed systems. A lot of "messing around" is needed, regardless of age, so don't be impatient with students who seem to want to do the same thing (in your eyes) over and over again. Encourage (but *don't* pressure) students to try to make up explanations for various behaviors that they notice in their closed systems. It is important to accept all explanations as worthy possibilities *and* to challenge all explanations to encourage continued thinking and activity.

Materials

Two sizes of test tubes, if possible, so that one easily fits inside the other, medicine droppers, clear plastic bottles (shampoo, liquid soap for dishes), rubber bands, "fat" balloons, plastic tubes used for separating golf clubs in the bag, stoppers to fit the tubes, old syringes without needles, rubber tubing, snap clothespins, water supply, lots of old newspapers, ice, hot plate or candles, large beaker.

Possible Activities

It may be worthwhile to make a couple of closed systems to help students understand the nature of a closed system. They can use this understanding as a starting point for their subsequent activities with closed systems.

1. Can you make a *closed system* like the one in sketch *a*?
2. Does anything change if the system is *heated* or *cooled*?
3. Does the height of the *water level* in the large tube affect the behavior of the system? In the small tube?
4. Can you make an *upside-down* closed system?
5. Can you make a *tall* closed system?

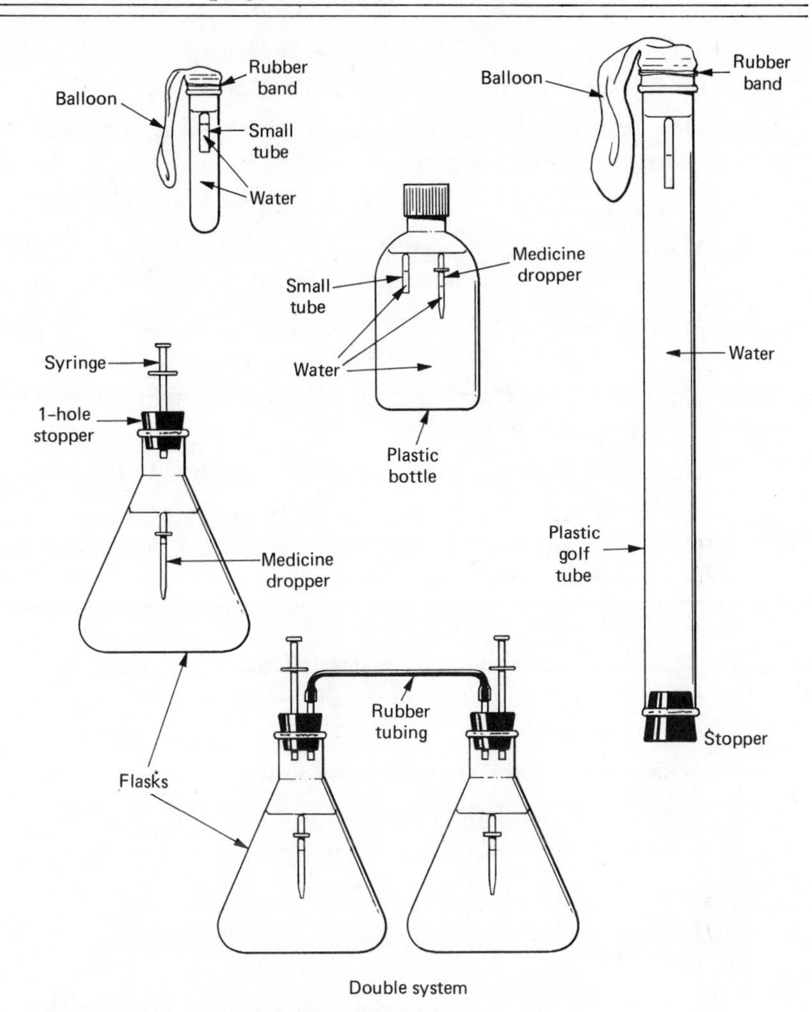

Figure 6–10. Materials for closed systems.

6. If a *medicine dropper* is used instead of a small test tube, does the system behave differently?
7. If two closed systems are *connected* with rubber tubing, how do they behave?
8. What happens if a "regular" closed system (sketch *a*) is *lowered* into a tall tube filled with water?
9. Can you construct a closed system using an empty *shampoo bottle* (plastic) and a medicine dropper (sketch *b*)?
10. Can you get the small tube (or dropper) inside a tall tube to *rise* once it has sunk to the bottom?

11. What seems to be the most important factor (variable) in determining whether the *inner tube* floats or sinks?
12. Does adding *liquid soap* to any of the closed systems have any effect on their behavior?
13. Can closed systems be constructed using *other materials?*
14. What happens if a "shampoo bottle with medicine dropper" system (sketch *b*) is placed in a *freezer* overnight?
15. What happens if closed systems (plastic type) are taken into a *swimming pool?*
16. How many closed systems can be connected together with *rubber tubing* and still work?
17. Can two tall plastic tubes be "stuck together" to form one *tall* closed system?
18. Can you get the small tube (inside the larger one) to go down about *halfway* and remain there?
19. How is a closed system like a *submarine?*
20. If a tall closed system is *suspended* by a long string from the ceiling and then swung in an arc, does the motion affect the system in any way?
21. Who can make the *fastest* "diver" in the class?
22. Does the *shape* of the large container (outside tube) affect the behavior of the small tube (medicine dropper) inside?
23. Can a *garbage pail* be used as a part of a closed system?
24. How many *medicine droppers* can be "sunk" in one, large closed system?
25. If a plastic closed system is suspended by a strong string and then *whirled* about, is the small tube affected?
26. If *cooking oil* is used instead of water, will the closed system behave the same?
27. What happens if *moth balls* are placed in a large closed system?
28. If *hot water* is used in a medicine dropper and then a closed system is made so that the dropper just barely hangs at the surface of the water, does anything happen as the water cools?
29. Will a closed system work if the medicine dropper is *inverted* so that the rubber part is down?
30. How can a *bicycle pump* be used as a part of a closed system?

Investigation Number Ten: Making Structures

If the names were important, this investigation could have been entitled "structural engineering" or "fundamentals in architecture." The simple title of making structures, however, more accurately describes the general intent of what most children will do with the various materials. There is no lower age limit suggested here as there was with the two previous investigations, since even the youngest students can make simple structures with commonly available materials. The development of perceptual and representational space in children is a very important part of the overall development of logical thought, as pointed out in Part Two of this book. All active movement in space by children facilitates this development, but this activity more than the previous ones helps children confront new problems involved with spatial relationships. One of the outdoor activities in the following section (about mapping) also tends to concentrate on spatial development as a means for problem solving.

Making structures can be a high interest, long-term investigation with many opportunities for social learning as well as conceptual learning. Since some of the structures may have to be built outdoors, be prepared to have the children working both indoors and out.

Materials

Cardboard boxes of all sizes, plastic soda straws, tape, toothpicks, scrap pieces of wood, nails and hammers, plastic wrap, spaghetti noodles (long and straight), pipe cleaners, paper for folding, art paper, scissors, glue, wire, plastic golf tubes (same as in closed system), other materials as needed.

Possible Activities

The size of the building material will determine the general size of the structure and, thus, whether the construction activity should be indoors or outdoors. Structures made from toothpicks, pipe cleaners, spaghetti noodles, soda straws, and other small materials can easily be built at a desk, table, or on the floor. Larger building materials may require outdoor activity by the students. Certainly, this will be one of the most active investigations of those described so far. Enjoy it!

1. What type of spaghetti noodle is best for making *tall* structures? Bridges?

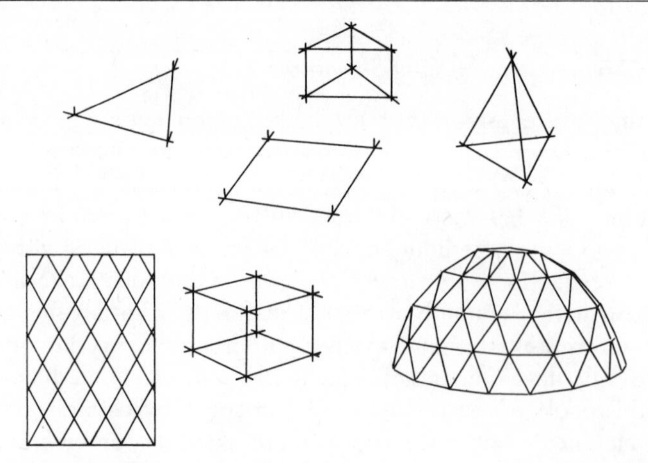

Figure 6–11. Example structures.

2. How many *different* structures can be made using 10 soda straws? Fifteen? Twenty?
3. Can you make a "geodesic" *dome* using soda straws and pins?
4. If only *triangles* are used as the basic building block, what kind of structures can be made?
5. If only *squares* are used as the basic building block, what kind of structures can be made?
6. Can *squares* and *triangles* be used together?
7. Who can make the *tallest* soda straw structure using the least number of straws?
8. What is the best way to make a *suspension* bridge?
9. How many different types of *bridges* can be constructed?
10. What kinds of structures can be made using *computer cards* only?
11. What materials could be used to construct a *water wheel*?
12. What types of *chairs* can be made using various materials?
13. Can a *chair* be constructed that will support someone's weight?
14. What types of structures can be made using only *five-sided* shapes (pentagons)? Six-sided shapes (hexagons)?
15. Why do you think a *beehive* is built of six-sided cells?
16. What types of *windmill* designs seem to work best?
17. How tall do you think a "*toothpick tower*" could be made?

18. What types of *mobiles* seem to move around best, when suspended from the ceiling?
19. What is the tallest tower that can be made using plastic *golf tubes* and tape?
20. Can you make a *balance* using only soda straws and straight pins?
21. Can you make an accordian-type of *gate* using only soda straws and pins?
22. Can you *visit* various construction projects in your town to see how the structures are built?
23. What seems to be the best way to make *round* (spherical) structures?
24. Can you construct a *box* that is "crushproof?"
25. Can you design a *city* with buildings, streets, and other structures?

outdoor science

Some activities are best done outside the confines of a school. There is a kind of unwritten code that causes most teachers, principals, parents, and even students to assume that really worthwhile learning can occur only inside a classroom. A classroom "closes in" and restricts what a person can do and learn whereas the outdoors, particularly if the school is surrounded with grass, bushes, and trees, expands the potential for learning. Most students can learn to accept responsibility for productive outdoor behavior quite readily, especially if they see that worthwhile learning experiences are involved. Some students may need more attention and help in learning to accept responsibility. However, these are the very individuals who can benefit most by outdoor science activities and, in fact, *need* opportunities for *self*-growth. Unless someone takes a *risk* with them, it is unlikely that they will develop more *trust* in themselves and in others.

The science activities in this final section are merely samples of the nearly unlimited possibilities for outdoor science. Though the physical setting does provide for certain limitations (for example lack of plants) the "under the sky" environment greatly extends the kinds of productive investigations that can be pursued. Though common sense is probably the best guide in developing a successful outdoor science component of your total program, a few suggestions are offered here as *possible* aids in extending your teaching resources:

1. Make plans *with* the students for the outdoor science.
2. As a group, identify *reasonable* limitations that the students should follow in their outdoor behavior.
3. Make it clear that disruptive behavior can endanger the possibility for continued outdoor science activities. Encourage the students to help one another remember the reasonable limitations agreed upon so that a few students do not ruin a good situation for the others.
4. *Always* be sure students take some type of activity-related material (measuring device, paper and pencil, collecting container) with them to use during the investigation. This seems to help students remember the purpose of going out-of-doors is for science not for playing tag, hide-and-seek, or other games.
5. If the activity suggests that students will range over a fairly wide area, have some means of signaling the end of the activity.
6. Try to schedule some time immediately after the students reassemble in the classroom to *continue* the activity. This may take the form of examining collected materials, sharing experiences in small groups, making records of some sort, or planning for the next day's science activity. If anything needs to be brought up before the entire class, this is a very good time to do it.

Investigation Number One: Bubbles, Balloons, and Paper Planes

Genuine learning always involves problem solving and the motivation to try to solve the problems (without coercion) involves interest. Problems must be seen as worthwhile, interesting, and not too difficult before we will attempt to search for possible solutions. Most students, even through the high school grades, rely heavily on interest rather than practical utility as motivation for engaging in problem solving. A dull, practical problem usually loses out to an interesting problem that has little or no apparent utility in one's life.

Blowing soap bubbles is certainly not seen by many people as a "critical life skill" but is a high interest activity as evidenced by the behaviors of individuals who are involved in blowing bubbles. The same is true of balloons and paper planes although planes somehow *seem* more like "science" and practical in terms of utility in one's life. All three systems combine high interest with ease of obtaining the necessary materials, especially the paper planes.

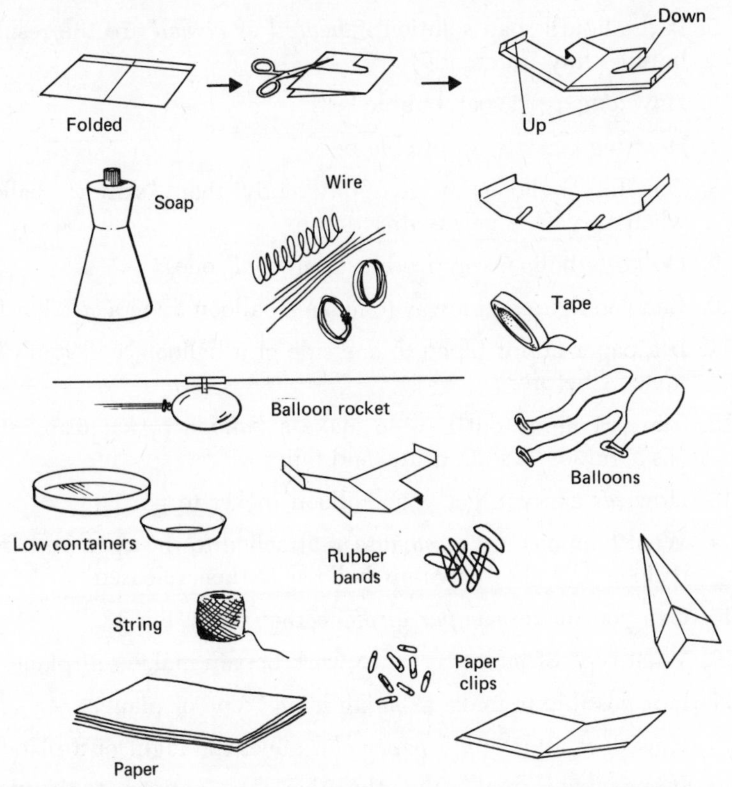

Figure 6–12. Materials for bubbles, balloons, and paper planes.

Materials

For soap bubbles—liquid soap, water, copper wire, wire cutter, containers for the soap solution; *for balloons*—balloons of all shapes and sizes, tape, string, rubber bands; *for paper planes*—paper of various types, sizes, and weights, tape, paper clips.

Possible Activities

1. What kind of soap *solution* seems to make the best bubbles?
2. Are some *brands* of liquid soap better than others for making bubbles?
3. If the copper wire is shaped into a *square,* can "squarish" bubbles be produced with it?
4. Do *different* shapes and sizes of copper wire *loops* make different shapes and sizes of bubbles?

5. If the liquid soap solution is *heated* or *cooled* are the resulting bubbles any different?
6. How *long* can a soap bubble last?
7. How *big* can a soap bubble be?
8. Do "fat" balloons behave differently than "skinny" balloons when the air is released?
9. Do *large* balloons fly farther than *small* ones?
10. Can you figure out a way to make a balloon fly in a *straight* line?
11. If a paper clip is taped to one side of a balloon, will it "fly" the same as before?
12. Can you figure out how to make a *balloon rocket* using string (as a guide), a soda straw, and tape?
13. How *far* can you get your balloon rocket to go?
14. What happens if a *parachute* is attached to the open end of the balloon and the blown-up balloon is then released?
15. Can you make a *paper airplane* that really flies?
16. What type of *paper* seems to work best in making airplanes?
17. Is it possible to make a "*flying wing*" type of plane?
18. How does attaching a *paper clip* affect the flight of a plane?
19. If one *wing* is smaller than the other, what happens to the plane's flight?
20. Can you build a plane that makes a *circle* and returns?
21. Can you build a plane that makes a *loop?*
22. What kind of design seems to be the best *glider?*
23. Does the *weight* of the paper affect the flight of the plane?
24. How *small* can paper planes be made and still fly rather than just "dropping down"?
25. How *large* can paper planes be made before they fail to glide well?
26. Is it possible to make a *cardboard* airplane fly like one made of paper?
27. Will a *round* plane fly?
28. How does the size of the *wing* affect the plane's flight?
29. Can a *jet plane* be made using balloons?
30. What would happen if a paper plane *contest* was organized in your school?

Investigation Number Two: Maps and Measures

Making maps and trying to "read" maps already prepared can be a highly interesting activity, one that offers a wide range of possibilities. Any map typically involves distance measurements, symbols, and a certain orientation. Until about the age of nine or ten years, most children have little facility with translating sizes, perspectives, and the like into a "map" of reality. Distance conservation is usually achieved around the age of seven or eight years and iteration using some "standard" is usually achieved around the age of eight or nine years, so attempts at quantitative measurements (linear) are not very successful until after this point in development. These limitations should be noted but not interpreted in a way to suggest that the younger child is unable to engage in productive map-making or map-reading activities. Jerome Bruner's statement regarding the teaching of anything to anyone was a bit ambitious but serves to point out the wide range of difficulties of many concepts, including mapping. Younger children can make maps, however simple, and they can make rough estimates of *relative* distances. The ability of the five- and six-year-old to represent space is primitive compared to the eight- and nine-year-old but nonetheless advanced over the four-year-old.

Activities should be selected that take these developmental differences into account. It is easier to map a small, well-known space that is immediately present than to attempt the same for larger, lesser-known spaces.

Materials

Old computer "printout" paper, if possible, string, rulers, metersticks, yardsticks, tape measures, large pieces of brown wrapping paper, crayons, directional compasses, maps of various types—aerial, county, coastal, road, or the like (old maps are usually available through a state agency).

Possible Activities

Before going outdoors for mapping activities, try mapping the classroom and/or other places inside the school.

1. Can you make a "map" of the *school grounds* using no measuring devices?
2. How could a length of *string* be used to help in mapping your school?

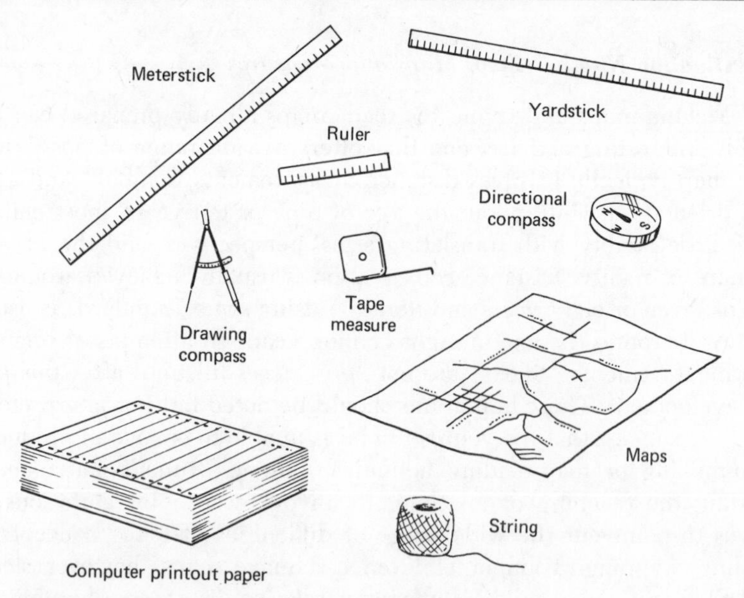

Figure 6–13. Materials for maps and measures.

3. Could you find a way of comparing the size of your classroom to the *cafeteria?*
4. If each student uses her or his *foot* as a measuring device, how do the lengths of various objects compare?
5. Who can come closest to guessing the length of a *car* in *feet?* A bus? An airplane?
6. How could the place where the *sun* rises be used as a part of a map?
7. If someone gave you a *partially completed* map of the school grounds, could you complete it?
8. Could you direct a classmate to an *exact spot* on the school grounds, using only written directions?
9. Could you make a map of a *neighborhood,* including streets, houses, and other structures?
10. Could you make up instructions for someone using a directional *compass* on how to go to a certain location?
11. If everyone in your class used the *same* measuring device (piece of string, meterstick), do you think all measurements would be the same?
12. What type of map of the school or school grounds could you make while being *blindfolded?*

13. Can *two* people make a map better than one?
14. Can you make a map of an *unknown* place?
15. If you let 1 inch (or centimeter) on your map be equal to one of your *steps,* how many inches (or centimeters) of paper are needed to map your classroom?
16. Does your *seat* in the classroom face north, south, east, or west?
17. Have the maps that you *have made* thus far faced north, south, east, or west?
18. If you are asked to make a map of the school using only one regular sheet of paper, how many of your steps (paces) will have to equal 1 inch (centimeter)?
19. How many rotations of an old *car tire* are needed to roll the whole length of your school?
20. Could you make a map using *tire rotations* as the basic unit of measurement?
21. What do you think your school and school grounds would look like from an *airplane?*
22. Can you locate your school on a *city map?* Your home?
23. Can you organize a *treasure hunt* using maps made by members of your class?
24. Can you make a map of the trip you take *going to school* each day?
25. How many rotations of the *tires* on your family's *car* are needed to go from your home to the school? Your bicycle?
26. If someone drove you around your town, do you think you could make a map of the *town* that could be used by another person?
27. What do you think is the *hardest* part of map making?
28. How many different *types* of maps might you find in the library?
29. How could you find out what type of work *map-makers* really do?
30. What do you think a map of the *moon* would look like?

Investigation Number Three: Just Watching

Albert Einstein's reference to science as a "refinement" of everyday thinking is certainly appropriate for this activity. Visual perception is a tremendously important sensory device, but it is quite different from *conception.* The experienced bird-watcher has developed the ability to observe with great care and patience. The animal psychologist develops a keen sensitivity to animal movements and to their meanings. The astronomer develops a capability to notice patterns in the night sky that are not readily apparent to the novice observer. The difference between the serious, experienced observer and the beginner is not perceptual but rather conceptual. By the time children are seven or eight years old they are usually *perceptually* capable of discriminating at an adult level, but their *conceptual* development is far behind the adult who is capable of formal or abstract thought. Thus, what we "see" in terms of patterns and relationships in our environment is heavily dependent upon our developed ability to conceptualize or make sense of these relationships.

"Just Watching" is intended as a source of opportunities for students to try to "refine" their usual observing behavior into a more careful, patient approach to "seeing" things. Static patterns as well as dynamic systems can be used as the "stuff" for watching, and all of the activities described thus far involve a certain amount of this stuff. The school's outdoor environment can be a limiting factor to what is available to watch, especially living things, but there are still many possibilities, even in the heart of a large city.

Materials

String, writing paper, art paper, pencils and crayons, audio tape recorder (if possible), Polaroid camera (if possible).

Possible Activities

1. How many different *living things* could you find outdoors in five minutes? Thirty minutes?
2. How many different kinds of *insects* live outside your school?
3. Could you write a *story* about how you think you would feel if you were a tree? A flower? A butterfly?
4. How many different *sounds* could you identify outside your school?
5. Is there evidence of *change* in things around your school?
6. What part of the school day seems to be the *quiestest* part?
7. How many *birds* can you find in 1 hour?
8. Do *squirrels* seem to be more active in the morning or in the afternoon?

Figure 6–14. Materials for just watching.

9. Can you figure out a way to find out if *plant parts* move during the day?
10. What do you think the *land* was like before your school was built?
12. Do cloud formations *change* during the day?
13. Do *shadows* change during the day?
14. Do any changes occur in plant life after a *rain*?
15. Do any changes occur in insect life after a *rain*?
16. Are there places around your school where *rain* collects?
17. Are there any *quiet* places around the school grounds?

18. Does the *wind* seem to blow more from any one direction?
19. Does the *wind* seem to blow more in the morning or the afternoon?
20. Do *birds* seem to prefer some trees (for sitting) over others?
21. How long does it take a *bird* to build a nest?
22. How long does it take a *spider* to spin a web?
23. Do all jet planes leave a *vapor trail* that can be seen?
24. Does there seem to be any pattern in how the *branches* on a tree are attached to the trunk?
25. Does there seem to be any pattern in how *leaves* grow out of the branches of trees?

Investigation Number Four: How Many and How Much

Children's abilities to quantify things in their environment varies widely, according to many factors. A particularly important ability in quantification is being able to estimate such things as weight, number, area, and volume of "pieces" of our environment. The need to find out how many or how much of something motivates us to try and quantify things. Children enjoy trying to find out how many and/or how much by measuring and guessing. An estimate is merely a guess with some reasonably valid basis. The ability to make accurate estimates increases according to developmental ability and with opportunities for such activity. Young children have restricted developmental abilities to all types of quantification and estimation, but nevertheless are capable of elementary types of such activity. Estimating that there are 100 of something can be just as difficult for the six-year-old as estimating 1000 of something for the nine-year-old. The basic ideas (questions) can be adapted according to the apparent abilities of the children involved.

Materials

Paper, scissors, jars of various shapes and sizes, balance (double pan), miscellaneous materials as needed.

Possible Activities

1. How many blades of *grass* can you cover with your hand? A dollar bill? A square inch of paper?
2. How many blades of grass are needed to equal the *weight* of a dollar bill? A cotton ball? A grain of corn?

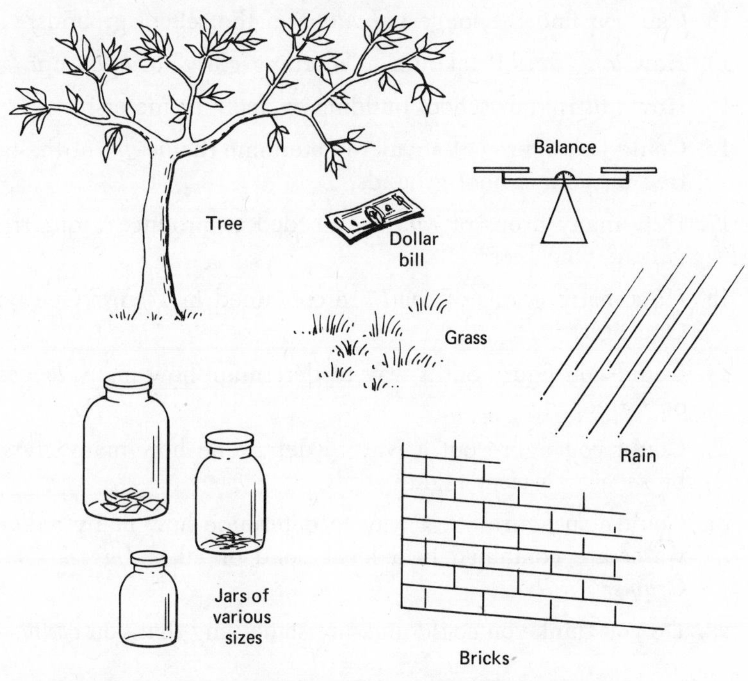

Figure 6–15. Materials for "how many" and "how much."

3. How many "ground up" dead *leaves* are needed to fill a drinking glass? A "cup"? A gallon container? A "liter"?
4. How many *leaves* are on a small tree? A large tree?
5. How many *nails* would be needed to build a bird house? A dog house? A people house?
6. How many *hairs* are there on a person's arm? Head? Leg?
7. How much *water* could be collected during a rain, using a nylon or plastic cover that is 2 feet on a side? Four feet? One meter?
8. How many *ants* are needed to equal the weight of a pencil? A toothpick? A drop of water? A cockroach?
9. How many "squirts" can be gotten from a *squirt gun* using a quart of water? A liter?
10. How many *bricks* are there in your school building? Windows?
11. How many *steps* are required to go completely around your school? Feet? Yards? Meters?
12. Can you find something that *weighs* 1 ounce? One pound? One gram? One kilogram?

13. Can you find the longest *shadow* on the school grounds?
14. How *long* does it take for a "morning glory" to open up?
15. How *tall* is your school building, in feet? In meters?
16. Could you figure out a way to determine the *height* of the tallest tree on your school grounds?
17. How many drops of *rain* are needed to produce 1 quart? One gallon? One liter?
18. How many grains of *sand* are contained in 1 cup? One quart? One liter?
19. Could you figure out a way to determine how many *leaves* are on a tree?
20. Could you figure out a way to determine how many *stars* can be seen on a clear night?
21. Could you figure out a way to determine how many *gallons* of water are contained in a local pond or lake? *Liters? Pounds? Grams?*
22. Do you think you could measure something that you *cannot see?*

Some Thoughts on Children, Science, and Teaching

The "possible activities" in this part of the book have been presented as ideas for helping children learn science. The nature of children and the nature of science both direct the teacher to view her or his role as establishing a classroom environment that facilitates personal investigation by children in ways that make sense to them. Helping children to become confident, independent problem solvers who continually ask questions is perhaps the most important gift that we can give to succeeding generations. Reading, 'riting, and 'rithmetic are certainly important skills to be learned and the schools should do their utmost to help all children become reasonably proficient in them, but science offers a unique opportunity for children to develop abilities for rational thought in a very concrete way. It is very likely that advanced technology will eventually eliminate the need for reading, writing, and arithmetic skills as we know them today. The current use of electronic calculators is an indicator that such will soon be the case for basic arithmetic skills and it is not hard to imagine an advanced technology that will make our normal reading methods obsolete. Rational thought, however, will always be important to the maintenance of a democratic society and to the preservation of our spaceship, Earth. Advanced technology must be monitored and improved by people who are capable of making rational decisions about the welfare of all peoples and their environment. A truly democratic society requires rational thought on the part of all the citizens if that society is to be maintained for the benefit of all the people.

Even for the four- and five-year-olds, science can be used as a vehicle to encourage personal exploration of meaningful problems that can make sense to the children. Their developmental level restricts the nature of the problems that can be dealt with, but nonetheless real problems can and should be a significant part of their schooling. Many of the ideas for investigations in the previous sections on Indoor and Outdoor Science can be used equally well with a four- to five-year-old or with an adult, depending on the *specific problem* that is conceptualized by the individual. Very young children can and do find out about insects and so do scientists. The nature of their "research" methods of finding out vary widely, according to respective developmental levels of logical thought. The most important ingredient for both the four- and five-year-old and the adult researcher is an environment in which they feel *safe* to pursue meaningful problems in ways that make sense to them. For the students in our schools, it is primarily the teacher who is in the best position to establish an environment that is "safe" for personal exploration of meaningful problems. *Safe* means being relatively free from threat so the teacher must try to remove such barriers as evaluation, especially for those students who have rather shaky concepts of their own abilities. It was emphasized in the ideas for indoor and outdoor science that no one question was more important than any other. Just as important, and perhaps more so, is the realization that *no one idea* offered by students is more important than another. The main purpose for investigations in science for children is to further their ability to investigate, not merely to know some result, although learning about results is a logical part of the process.

Teaching science as suggested by this book requires a teacher who understands children and wants to help them further develop their interests and abilities in finding out about things in their environment. Allowing children the freedom necessary to develop responsible, self-directed problem-solving behavior requires a professional teacher who is confident enough in his or her abilities to take a risk. It *is* a little more risky to allow children to pursue problems that they see as meaningful, in ways that make sense to them. The extent to which you as a teacher are willing to take a few risks with your students by allowing increased intellectual freedom will be tied directly to your real commitment to such goals for students as the following:

1. Greater tendency for and confidence in finding out.
2. Increased curiosity for how things happen.

3. Acceptance of "mistakes" as valuable information in further exploration.
4. Greater tendency to "question the experts" rather than accepting things unquestioningly.
5. Increased appreciation for the beauty in nature.
6. Increased commitment to taking action when rational thought requires such action (for example, preserving the natural environment and preserving constitutional freedoms).
7. Greater acceptance of responsibility for one's own actions.

What you do as a teacher can and will make a difference!

appendix a

A Five-Year Study of the Development and Implementation of Two Quantitatively Defined Teaching Strategies for Elementary and Secondary School Science:

STUDENT-STRUCTURED LEARNING IN SCIENCE (SSLS)
and
TEACHER-STRUCTURED LEARNING IN SCIENCE (TSLS)*

* These two studies by Charles C. Matthews, Ronald Good, James Shymansky, Patricia Kolebas, John Penick, and Tom Allen were presented to The National Association for Research in Science Teaching at the Forty-Seventh Annual Convention Chicago, Illinois, Sheraton-Blackstone Hotel April 15–18, 1974. This paper reports research that is a part of Project LEO—Studies of Learning Environments and Outcomes, Florida State University. It is printed here with permission of the authors' collaborators.

introduction

In introducing his book, *The Hidden Persuaders,* Vance Packard wrote the following in 1957:

> It is about the large-scale efforts being made, often with impressive success, to channel our unthinking habits . . . and our thought processes by the use of insights gleaned from psychiatry and the social sciences. Typically these efforts take place beneath our level of awareness; so that the appeals which move us are often, in a sense, hidden. The result is that many of us are being influenced and manipulated far more than we realize, in the patterns of our everyday lives.

Of course Packard was writing about behaviorism as it applies to massive advertising campaigns to create markets for products for which there exists no consumer demand. Many are concerned with the increasing use of behavior management techniques for the purpose of manipulating young people who have not yet had the opportunity of developing their own unique personalities, interests, and needs. Critics of behaviorism make the assumption that the following 1957 statement by Packard applies in 1974 to treatment of young people in schools:

> Some of the manipulating being attempted is simply amusing. Some of it is disquieting, particularly when viewed as a portent of what may be ahead on a more intensive and effective scale for us all. Co-operative scientists have come along providentially to furnish some awesome tools.

This paper describes briefly the 1968–74 activities of two related research and development projects at Florida State University. The CSLS Project (Child-Structured Learning in Science) of 1968 has gradually evolved into Project LEO (Learning Environments and Outcomes) of 1974. The CSLS Project developed humanistic instructional strategies and Project LEO has developed quantitative definitions for these strategies and for contrasting behavioristic strategies. Project LEO is now studying differences in environments and outcomes associated with two contrasting strategies identified as student-structured learning in science (SSLS) and teacher-structured learning in science (TSLS). SSLS strategies are viewed by the investigators as humanistic and TSLS strategies are viewed as behavioristic. Project LEO takes the philosophical argument between humanists and behaviorists into the laboratory.

child-structured learning in science, 1968–1972

The CSLS Project was an effort to identify and implement activities and materials that would facilitate maximum affective and cognitive learning in science. The project began in 1968 with initial funding from the Georgia Department of Education and the National Instructional Television Center. Funding was subsequently received from Florida State University and the Florida Department of Education. Much of the effort that went into this project, however, consisted of donations of time and expertise by psychologists, scientists, science educators, students and teachers in many parts of the United States and Canada.

The CSLS Project reflected the following assumptions (1, 2):

1. It is possible to logically derive learning conditions from goals and characteristics of learners.
2. Learning conditions must reflect what is *not* known as well as what is known about students.
3. Teacher behavior and learning materials are the dominant factors in determining learning conditions—since these two factors communicate to the students both conceptual and operational meanings of learning.
4. Learning how to learn can be facilitated by school experiences.
5. Self-actualized learning is the most important goal of education.
6. Learning conditions can be tested by studying interactive processes and outcomes of educational activities.

CSLS Goals

The broad goals of CSLS activities are outgrowths of value judgments made by several hundred scientists, science educators, human development specialists, teachers, parents, and students. These goals can be characterized as "humanistic" in intent because they (1) focus on the development of the individual's full potential, (2) facilitate individuality in learners, and (3) are associated with productive learning under conditions of maximum freedom.

CSLS goals emphasize conceptual learning over skill learning. According to CSLS, skill learning can result from imitation (verbal or nonverbal) but conceptual learning grows out of the cognition and affect of the student. Conceptual learning is that learning which is created or structured within and by the student. All CSLS activities have the following long-term goals:

1. To enhance systematic and creative thinking abilities of students in pursuit of investigations or solutions to problems.
2. To enhance students' beliefs that they can interpret and manipulate their own environments—that they are part of their environments and dependent upon them.
3. To facilitate for students the development of positive self-concepts with regard to independent learning and responsible manipulation of their environments.
4. To facilitate individual development of interests, attitudes, personalities, and creativity in order to enchance the continued development of individuality in students.
5. To facilitate students' tendencies to accept other individuals—especially those who have ideas and values which are different from their own.

These long-term conceptual goals are operationally defined by more specific statements of desired affective and cognitive learning. CSLS affective objectives are associated with the development of positive self-concept with regard to independent learning. Within the CSLS Project, this means the following:

1. Students identify themselves as persons who can successfully study their environments, and who choose to use what they learn about their environments.
2. Students describe learning in terms of activities which make sense to them and they feel comfortable with these descriptions.
3. Students comfortably accept their own explanations for natural phenomena and modify these only when they cease to be compatible with their own interpretations of their environments.
4. Students feel comfortable in stating alternative explanations for observed phenomena.
5. Students identify and accept tentativeness as an important characteristic of knowledge.

CSLS cognitive objectives are associated with the goal of communicating how creative and systematic thinking relate to solving self-perceived problems. Students should be able to design activities without prompting and complete, without instruction, activities in which they do the following:

1. Students manipulate objects or systems in ways which are dependent upon self-perceived properties of the objects or systems.

2. Students identify relationships among properties of objects or among the factors which affect the behaviors of systems.
3. Students manipulate objects and systems to test the usefulness of the relationships which they have identified.

CSLS Research Base

Although CSLS activities obviously take into account physical characteristics of learners, the research base for CSLS activities is associated primarily with emotional and cognitive characteristics of learners. The work of numerous researchers has shed considerable light on the emotional characteristics that people have in common. The following research-based statements have relevance to CSLS activities:

1. The human being is characterized by a tendency toward learning. Unless barriers have been erected, a person will choose to learn (3).
2. "Need to know" is insatiable and continually forcing persons to rearrange ideas into patterns that make more sense to them (4).

Cognitive characteristics of students are those characteristics associated with the thinking that is available to students. The Piagetian school, Bruner's group, and certain of the Soviet psychologists envision these characteristics as representation systems available to individuals for dealing with future encounters of the organism with reality (5, 6, 7). A brief listing of research supported statements on cognitive characteristics that grew out of Piaget's work is given below (5):

1. The cognition of human beings progresses through stages characterized by increasingly powerful representation systems.
2. The ordering of these stages of cognitive development is constant and has been found in all societies studied.
3. Chronological ages associated with stages of cognitive development vary from one person to another.
4. Interpretation of the environment and problem solving associated with this interpretation are limited by representation systems available to individuals.

CSLS takes into account the following well-known examples of research-supported statements associated with facilitative classroom behavioral patterns:

1. Openness and self-directiveness characterize environments which tend to facilitate goals associated with creativity in the sciences and wisdom in those fields with social concern (8).
2. Students tend to copy the behavioral patterns exhibited by the teacher. If the teacher tries to dominate students, students try to dominate

each other; if the teacher accepts students, students tend to be accepting of each other (9, 10, 11).

3. Directive teaching behaviors tend to produce disruptive anxieties in students and reduce the learning of new concepts (12, 13).
4. Constraint teacher behaviors tend to produce a high level of dependency of students on their teacher (14).
5. Directing students to engage in manipulative or verbal operations which they cannot engage in mentally tends to erect knowledge superstructures which crumble under the slightest cognitive stress (15).
6. Long-term learning seems to be characterized by personal involvement, self-initiation, a sense of discovery, pervasiveness, self-evaluation, and meaningfulness (3).

In identifying CSLS materials, the following two research-based statements were heeded.

1. Knowing an object requires acting on the object—modifying it, transforming it. Learning seems to result from what one does to objects in the environment and the doing must be both physical and mental (5).
2. Logical thinking does not derive from verbal learning but rather, from a total coordination of actions on object (5).

This led CSLS to distinguish between representational and nonrepresentational materials and to place almost total dependence on nonrepresentational materials—until the student initiated an interest in the use of representational materials. Objects were used for themselves rather than as representations of other objects, events, or ideas. A blue glass marble, as a nonrepresentational object, is whatever the student perceives it to be. It is not presented as a representation of glass objects, blue objects, spherical objects, or any other class of objects. Obviously it is not used to represent an atom, a nucleus, or a molecule. Printed or pictorial materials, as representational materials, give way to manipulative, nonrepresentational materials. Words are used but never substituted for objects or events by CSLS teachers. Objects are available for direct manipulation by the student—and the object is what the student perceives it to be at any moment. Student manipulations mean what students perceive them to mean.

CSLS Learning Conditions

A major activity of the CSLS project was the derivation from goals and student characteristics of a set of learning conditions. CSLS classroom conditions, which grew out of the previously stated goals and research on student characteristics, are characterized by the following:

1. Students must have access to a variety of materials which they are free to use or ignore in an environment that allows, but does not require, interaction with the teacher and/or other students.
2. The teacher cannot know students' precise cognitive levels and therefore cannot prescribe what they are to learn or how they are to learn.
3. A CSLC session has no specific beginning, middle, or end.
4. Students feel that they can do what they wish with the materials so long as they do not disturb other students or damage the materials unnecessarily.
5. Students terminate sessions when they wish; this decision is made by each individual.
6. Students work individually or in groups as they choose, and this decision is made by each student at any time.

The role of the teacher and materials provided by the teacher are described generally by the following six statements.

1. The teacher makes available the greatest possible variety of materials for students to use.
2. The teacher responds to what students are doing rather than giving directions for students to follow. Printed materials do not prescribe or give directions.
3. The teacher responds to *individual* students rather than giving a generalized response to groups of students.
4. The teacher responds by asking questions or making neutral (non-evaluative) comments, to which students may respond or not as they choose.
5. The teacher does not reward or evaluate students for their activities. To reward or evaluate would communicate to students that the teacher wants them to discover the secret objectives of the lesson.
6. The teacher accepts (but does not reinforce) both "correct" and "incorrect" statements by the students. To do otherwise would suggest that the teacher knows what the student's perception is better than the student knows it. Neither does the teacher provide printed feedback for the purpose of establishing "correctness" of the student's perceptions.

Early CSLS Comparative Studies

Investigations (16, 17) conducted during the 1968–69 school year compared three different science programs for children within the age range five and one-half to six and one-half years. Nineteen teachers and

570 students were studied for the purpose of comparing the following: A textbook-based program, *Science—A Process Approach Part A.* and *CSLS Level One* (1, 2).

Utilizing the Science Curriculum Assessment System (18), teacher behaviors, student behaviors, and student cognitive developmental characteristics were studied. Examination of the data from these investigations made it obvious that *CSLS Level One* involved decidedly less teacher directiveness and decidedly more teacher interaction with individuals and small groups of students. The study also made it clear that *CSLS Level One* students did more self-initiated and self-designed activities with no increase in disruptive activity or idleness. Data on cognitive developmental levels of students revealed no significant differences among programs and placed the majority of students at Piaget's preoperational level.

project leo, 1971–1974

In September 1971, a project was initiated to study learning environments and outcomes associated with the strategies developed previously in CSLS. These were to be compared with a contrasting set of strategies that had been observed to be in widespread usage for K–12 science teaching. The two sets of strategies were identified as "student-structured learning in science" (SSLS) and "teacher-structured learning in science" (TSLS). Although these strategies are defined quantitatively in terms of contrasting teaching behaviors, the overall framework for the study is better understood in terms of contrasts between SSLS and TSLS goals and activities—as well as teacher behaviors.

SSLS and TSLS Goals

SSLS goals are identical to those given previously as CSLS Goals on page 3. TSLS goals are derived from the practices of numerous K–12 science teachers and reflect what the Project LEO investigators consider important contrasts with SSLS goals. Whereas SSLS goals emphasize conceptual learning and an open (humanistic) environment, TSLS goals emphasize skill learning and a closed (behavioristic) environment. When implementing TSLS strategies, the teacher investigators of Project LEO generated behaviors from a framework in which *learning selected skills was of major importance and training the student in these skills was a major goal.* More specifically, TSLS long-term goals are the following:

1. To enhance the verbal and manipulative skills of students in the pursuit of activities compatible with the structure and processes of science and the competencies of the student—as perceived by the teacher or other experts.
2. To enhance students' beliefs that their careful attention to directions based on the judgment of authorities will preserve their environments.
3. To facilitate for students the development of positive self-concepts with regard to following directions in the manipulation of their environments.
4. To facilitate the development of interests, attitudes, and personalities which fit into the expert's perception of societal needs.
5. To facilitate students' tendencies to seek ideas and values that can remain unchanged throughout their lives.

These long-term, skill-oriented goals are operationally defined by more specific statements of desired affective and cognitive learning. TSLS affective objectives are associated with the development of an accurate concept of self with regard to skill learning. Within the TSLS framework this means the following:

1. Students correctly assess and accept their own science skills relative to those of their peers.
2. Students accept science as overt verbal and manipulative skills and correct knowledge.
3. Students feel comfortable in stating what experts recognize as correct explanations for natural phenomena that they have covered in their science program.
4. Students feel comfortable in stating the explanations (for observed phenomena) which have been told previously to them by the teacher or another authority.
5. Students identify and accept correctness as the most important characteristic of scientific knowledge.

TSLS cognitive objectives are associated with training students in the skills of science and how these skills are used in efficiently and correctly solving practical problems. Students who complete the TSLS program should be able to follow instructions that require that they do the following:

1. Students manipulate objects in ways which correctly identify and measure the properties of the objects.
2. Students correctly identify relationships among properties of objects or among the factors which affect the behaviors of systems.

3. Students use expert sources to check the correctness of properties of objects, the relationships which they have identified, and general knowledge.

TSLS and SSLS Teacher Behaviors

Utilizing a modification of teacher behavior categories developed by Matthews (18) in 1968, SSLS and TSLS teacher behaviors are defined quantitatively in Table 1. It is obvious that SSLS teacher behaviors

TABLE 1. SCAS Teacher Behavior* Ranges for SSLS and TSLS

Verbal and Nonverbal Teacher Behaviors	SSLS Per Cent	TSLS Per Cent
1. Does not observe student behavior.	0–1	0–1
2. Observes student behavior but does not respond.	40–60	20–40
3. Accepts (without evaluation) student behavior.	10–20	1–2
4. Praises or positively evaluates student behavior.	0–2	10–20
5. Rejects and/or discourages student behavior.	0–2	0–20
6. Reprimands student, criticizes ridicules, uses sarcasm.	0–1	0–1
7. Asks question which does not tell the student what to do or how to do an activity.	20–40	15–30
8. Gives information which does not tell the student what to do or how to do an activity.	1–5	1–5
9. Makes statement or asks question which tells the student what to do or how to do an activity.	0–2	20–40

* SCAS teacher behavior categories also distinguish between teacher interaction with 1–6 students ("S") and teacher interaction with 7 or more students ("T"). Less than 1 per cent of SSLS and TSLS teacher behaviors were "T" behaviors. For category definitions, see *Handbook for the Application of the Science Curriculum Assessment System* (18). Note that categories 3, 4, and 9 have been modified.

in Table 1 are essentially responsive (nondirective) and neutrally accepting (nonevaluative) toward students. Both SSLS and TSLS teacher behaviors include questions and observations.

An important condition for SSLS and TSLS that is not reflected in Table 1 is the number of students with which the teacher interacts at a given moment. Although group size for both SSLS and TSLS was 25–30 students, 99 per cent of teacher behaviors involved interaction with fewer than seven students—usually one or two students. Therefore, the teacher was essentially "roving" among students; the low level of behavior in SCAS category 1 indicates the extent to which teachers remained active during SSLS and TSLS science. Both SSLS and TSLS are "individualized science," TSLS being prescriptive and SSLS being nonprescriptive. TSLS teaching narrows alternatives for students to those which the teacher identifies as most efficient for the student; SSLS teaching maximizes alternatives (short of endangering the rights of other people).

SSLS and TSLS Activities

In studying the effects of contrasting teacher behaviors on students, all other conditions were held constant. These included materials and other physical facilities. In order to further clarify SSLS and TSLS learning conditions, however, contrasting descriptions of activities are given in Table 2. Since all materials for SSLS and TSLS are identical, the differences in the activities result from the differences in the teacher behaviors shown in Table 1.

The SSLS versus TSLS Investigation—Overview

During the summer of 1971, quantitative definitions of SSLS and TSLS teaching behaviors were finalized and the teacher learned sets of contrasting behaviors and demonstrated these consistently within the ranges shown in Table 1. Activities were identified, materials were collected, and pre- and posttest instruments were developed. Research assistants were trained for data collection.

Beginning in September 1971, a pilot study was conducted with 52 students ranging from nine and one-half years to ten and one-half years of age. Twenty-six students were randomly assigned to SSLS and the remaining twenty-six constituted the TSLS group. Students were assigned with a class-ranking restriction on randomization to insure a range of students in each section. Using the *California Short-Form Test of Mental Maturity* and fourth-grade achievement grades, students were privately designated as high-rank, middle-rank, and low-rank. During

TABLE 2. General Description of SSLS and TSLS Activities*

Activity Type	SSLS	TSLS
One	Students use self-selected sets of nonrepresentational materials (which are usually provided by the teacher) for the creation of structures and patterns and for fantasy and game-playing. Students (at any cognitive level) who have had limited experience with independent investigation usually engage in Type One Activities. Students at the preoperational level of cognition (whether experienced or not) usually engage in Type One Activities. *See CSLS Level One* (1, 2) for examples.	Students use materials (as directed by teacher) for engaging in the identification and measurement of specific predetermined properties of objects. In addition to observing and measuring, students use materials for classifying, counting, pretending, and stating their findings. When the teacher considers it appropriate, the child moves to Type Two Activity.
Two	Students use self-selected sets of nonrepresentational materials for seeking relationships among the properties of objects. Without pressure to get "correct" relationships students are given opportunities to identify and make statements of relationships in any way they choose. Much variation from one student to another is observed and accepted. Inexperienced students at the formal operational level usually choose this type of activity after some experience in Type One Activity. Both experienced and inexperienced students at the concrete operational level usually choose this type of activity–again after some experience in Type One Activity. See *CSLS Level One* (1, 2) for examples.	Students use materials (as directed by the teacher) for making predictions and pictorial representations and for identifying specific predetermined relationships among the properties of the objects. When the teacher considers it appropriate, the student moves back to Type One Activity or on to Type Three Activity.
Three	Students use self-selected nonrepresentational materials for seeking relationships among the various factors associated with the behavior of systems. Students usually make gradual shifts through Type One and Type Two to Type Three Activities as their cognitive development and experience permit. Experienced students at the late concrete operational stage and formal operational stage usually choose Type Three Activity. See SSLS (19, 20) for examples.	Students use materials (as directed by the teacher) for identifying properties of objects and relationships among the factors which influence the behaviors of systems. Students follow instructions on inferring, defining operationally, formulating hypotheses, and interpreting data. Students regress to Type One or Type Two or progress to Type Four at the discretion of the teacher.

* A subsequent publication will describe specific SSLS and TSLS activities of the five activity types given here.

TABLE 2 (*continued*)

Activity Type	SSLS	TSLS
Four	Students identify problems, materials and procedures and conduct investigations of their own design. Sets of objects are *not* presented to students by the teacher. Total freedom is given to students in identifying problems which interest them. (Obvious limitations of nonavailability of materials and personal safety are involved). Students functioning at the formal operational level and experienced in Type Three Activity usually choose Type Four Activity. See *SSLS* (19, 20) for examples.	With greater emphasis on representational materials, students follow the teacher's directions in the use of materials for identifying and controlling variables, experimenting, and formulating models. Students identify necessary materials for carrying out the teacher's instructions. At the teacher's discretion students usually proceed to Type Five Activity.
Five	Students utilize representational materials in investigating the processes and outcomes of workers in fields of their own interests, usually reading historical and/or current accounts of scientific work. Students who recognize that knowledge is person-made and tentative usually choose Type Five Activity without the displacement of the previous four types of activity. Students read about both success and failure in efforts of scholars and researchers. Type Five Activity is usually selected by widely experienced students who have reached formal operational cognition and who have a continuing interest in the field of endeavor under study.	Using representational materials, students follow the teacher's directions in using expert sources for the purpose of checking the correctness of their own acquisition of new information and for identifying procedures to use in their investigations. Based primarily on ideas from their readings, students identify materials and procedures for new investigations. With the teacher's approval, they conduct investigations which were not initiated by the teacher. At the teacher's discretion, students return to the previous four types of activity or continue with Type Five Activity.

daily SSLS and TSLS sessions students were observed individually and their behaviors were coded on a modification of SCAS Student Behavior Categories (18), which are given in Table 3. Utilizing Forms A and B of the *TAB Inventory of Science Processes* students were pre- and post-tested. These two studies were consistent in findings, yielding the following conclusions (21, 22):

1. SSLS low- and middle-ranked students did as well as SSLS high-ranked students in problem solving.
2. TSLS high-ranked students did about the same as SSLS high-ranked students in problem solving.
3. TSLS low-ranked students did not do as well as TSLS middle-ranked students and TSLS middle-ranked students scored lower than TSLS high-ranked students on problem solving.
4. When SCAS behaviors of SSLS and TSLS students were compared the TSLS students:

    a. did more "observing the teacher" (L1).

    b. did more "following teacher directions" (L2).

TABLE 3. SCAS Student Behavior Categories*

Lesson-Related (L)	Category Description	Non-Lesson-Related (N)
L1	Observes teacher or student acting for the teacher.	N1
L2	Follows teacher's directions as to what activity should be done and/or how the activity should be done.	N2
L3	Does not follow teacher direction as to what activity should be done and/or how the activity should be done; does activity of own design.	N3
L4	Responds to teacher question or request by telling or showing.	N4
L5	Initiates (or attempts to initiate) interaction with teacher; continues self-initiated interaction with teacher.	N5
L6	Initiates interaction with another student.	N6
L7	Receives ideas from another student (who is not acting for the teacher).	N7
L8	Copies (or follows directions of) another student—must be preceeded by 7.	N8
L9	Gives idea to another student (not at request of teacher).	N9

* For category descriptions see *Handbook for the Application of SCAS* (18). Note that categories 2 and 3 have been modified.

c. did less "self-designed activity" (L3).

d. did more "initiating interaction with teacher" (L5), and

e. did about the same amount of "lesson-related behaviors" (L) as did SSLS students.

5. SCAS BEHAVIORS OF TSLS low-ranked students were different from other TSLS students in that the low-ranked students:

a. exhibited less "lesson-related" behaviors (L),

b. "observed the teacher" (L1) more, and

c. "followed teacher directions" (L2) less.

6. SCAS behaviors of SSLS low-, middle-, and high-ranked students were about the same.

Based on these findings and utilizing new or modified instruments, a 1972–73 study was designed and implemented with 250 students ranging in age from five and one-half to ten and one-half years and with 50 high school students enrolled in first-year general chemistry. During the 30-week SSLS and TSLS treatment, students were studied in the following ways:

1. *Classroom behaviors.* Utilizing SCAS student behavior categories, students were observed individually and behaviors were coded (23).
2. *Problem-solving ability and confidence.* Utilizing an individual interview, students were pre- and posttested for ability and confidence in the solution to ten selected problems (24).
3. *Cognitive developmental characteristics.* Utilizing an individual interview students were pre- and posttested on conservation and similar tasks (25).
4. *Self-concept with regard to science and independent activity.* Utilizing a group-administered instrument (with individual assistance and monitoring), students were pre- and posttested on their feelings and tendencies regarding science and other independent activities (26).
5. *Information acquired.* Utilizing a set of standardized tests students were pre- and posttested on science information acquisition and information acquisition in general (27).
6. *Teacher impressions.* Teacher's made subjective judgments and kept daily logs of what they considered significant events (28). Participating teachers and investigators met once each week to discuss these and to plan for the next week.

Detailed findings of the 1972–73 investigations are given elsewhere (23, 24, 25, 26, 27, 28, 29), but it is possible to summarize briefly several conclusions based on these findings. Findings indicated that, after one school year of SSLS and TSLS conditions, SSLS students exhibited higher levels of

1. Ability to solve certain problems.
2. Confidence in their ability to solve certain problems.
3. Self-initiative and self-directiveness in engaging in learning activities.
4. Verbal creativity.

TSLS students exhibited higher levels of

1. Listening to and watching the teacher,
2. Carrying out teacher-designed activities,
3. Copying other students,
4. Initiating interaction with the teacher.

In general, it can be stated that the investigations conducted thus far have consistently indicated that *students are more independent and fully functioning under SSLS conditions than under TSLS conditions with no increase in disruptive behaviors.*

The 1973–74 school year is devoted to a second-year study of 52 of the elementary school students who were studied during 1972–73. These 52 students were retained in their original groupings and are currently continuing under SSLS and TSLS conditions. Additionally, two new groups of high school chemistry students are being studied—one group under SSLS and one group under TSLS conditions.

FUTURE PROJECT LEO ACTIVITIES*

The 1974–75 school year will be devoted to comparative studies of more subtle contrasting strategies within the SSLS framework. In addition to continuing the investigations of learning and outcomes, Project LEO will:

1. Provide consultative services to groups who wish to conduct their own research or who wish to implement SSLS- or TSLS-related teaching strategies and
2. Publish instruments, techniques, findings, and implications of CSLS and Project LEO investigations.

references

1. MATTHEWS, CHARLES C. ET AL. *Handbook for the CSLS Level One Teacher.* Tallahassee: Florida State University, 1969.
———. *Handbook for the CSLS Coordinator.* Tallahassee: Florida State University, 1969.
2. ROGERS, CARL. *Freedom to Learn.* Columbus, Ohio: Charles E. Merrill Publishing Company, 1969.

* Results from continuing LEO studies can be found in 1974–1976 issues of *The Journal of Research in Science Teaching* and *The Science Teacher* as well as in dissertations in science education from The Florida State University.

3. KAGAN, JEROME. *Understanding Children.* New York: Harcourt Brace Jovanovich, 1971.
4. PIAGET, JEAN. "Development and Learning," *Journal of Research in Science Teaching,* Vol. II, No. 3, 1964.
5. BRUNER, JEROME S., ET AL. *Studies in Cognitive Growth: A Collaboration at the Center for Cognitive Growth.* New York: John Wiley & Sons, 1966.
6. ZAPOROZHETS, A. V. "The Development of Perception in the Pre-School Child," *European Research in Cognitive Development, Monograph of the Society for Child Development.* Ed. by Paul H. Mussen. University of Chicago Press, Chicago: Vol. 30, No. 2, 1965.
7. SMITH, B. O. "Toward a Theory of Teaching," *Theory and Research in Teaching.* Ed. by Arno A. Bellack), New York: Bureau of Publications, Teachers College, Columbia University, 1963.
8. ANDERSON, HAROLD H., AND BREWER, HELEN M. "Studies of Teachers' Classroom Personality: Dominative and Socially Integrative Behavior of Kindergarten Teachers," *Applied Psychology Monograph,* No. 6, 1945.
9. ANDERSON, HAROLD H. AND BREWER, JOSEPH E. "Studies of Classroom Personalities: Effects of Teachers' Dominative and Integrative Contact on Children's Behavior," *Applied Psychology Monograph,* No. 8, 1946.
10. ANDERSON, HAROLD H., ET AL. "Studies of Teachers' Classroom Personalities: Follow-up Studies of the Effect of Dominative and Integrative Contact on Children's Behavior." *Applied Psychology Monograph.* No. 11, 1946.
11. FLANDERS, NED A. "Personal-Social Anxieties as a Factor in Experimental Learning Situations," *Journal of Educational Research,* Vol. 45 (Oct. 1951).
12. COGAN, N. L. "Theory and Design of a Study of Teacher Intraction," *Harvard Education Review,* Vol. 26, No. 4 (Fall 1956).
13. LEWIN, KURT, *Dynamic Theory of Personality.* New York, McGraw-Hill Book Company, 1935.
14. ALMY, MILLIE. *Young Children's Thinking.* New York: Teachers College Press, 1966.
15. MATTHEWS, C. C. ET AL. "CSLS: A Research and Development Project of the Department of Science Education, FSU," *Mirrors for Behavior: An Anthology of Observation Instruments,* Ed. by Simon and Boyer. Philadelphia, Research for Better Schools, 1970, Vol. XV.
16. MATTHEWS, C. AND PHILLIPS, D. "EIE Elementary School Science Project, Evaluation Report," Florida State University: Tallahassee, 1969.
17. MATTHEWS, CHARLES C., AND PHILLIPS, DARRELL G. *Handbook for the Application of the Science Curriculum Assessment System.* Tallahassee: Florida State University, 1968.
18. MATTHEWS, CHARLES C. ET AL. *Teacher's Guide for Student-Structured Learning in Science.* Dubuque, Iowa: W. C. Brown Publishers, 1971.
19. ———. *Student-Structured Learning in Science: A Program for the Ele-*

mentary School Teacher. Dubuque, Iowa: W. C. Brown Publishers, 1971.

20. SHYMANSKY, J. AND MATTHEWS, C. "A Comparative Laboratory Study of the Effects of Two Teaching Patterns on Certain Aspects of the Behavior of Students in Fifth Grade Science," Presented to the 46th annual NARST Convention, Detroit, 1973.
21. PENICK, J. ET AL. "A Comparative Study of the Effects of Teacher Behavior on Student Behavior in Fifth Grade Science: A Replication Study," Presented to the 46th annual NARST convention, Detroit, 1973.
22. MATTHEWS, C. ET AL. "A One-Year Study of Comparative Effects of SSLS and TSLS on the Student's Classroom Behaviors," Presented to the 47th annual NARST convention, Chicago, 1974.
23. ———. "A One-Year Study of Comparative Effects of SSLS and TSLS on the Student's Problem-Solving Ability and Confidence," Presented to the 47th annual NARST convention, Chicago, 1974.
24. GOOD, R. ET AL. "A One-Year Study of Comparative Effects of SSLS and TSLS on the Student's Cognitive Developmental Level." Presented to the 47th annual NARST convention, Chicago, 1974.
25. SHYMANSKY, J. ET AL. "A One-Year Study of Comparative Effects of SSLS and TSLS on the Student's Self-Perception with Regard to Science," Presented to the 47th annual NARST convention, Chicago, 1974.
26. MATTHEWS, C. ET AL. "A One-Year Study of Comparative Effects of SSLS and TSLS on the Student's Information Acquisition," Presented to the 47th annual NARST convention, Chicago, 1974.
27. LEONARD, J. ET AL. "Teachers' Perceptions of Comparative Effects of SSLS and TSLS on Elementary and Secondary School Students," Presented to the 47th annual NARST convention, Chicago, 1974.
28. KOLEBAS, P. ET AL. "A One-Year Preliminary Study of Comparative Effects of SSLS and TSLS on the Student's Behavior and on Affective and Cognitive Learning in Secondary Chemistry," Presented to the 47th annual NARST convention, Chicago, 1974.

appendix b

The Traditional Sequencing of Mathematics for Young Children Should Be Changed

by Ronald G. Good

* This paper appeared in *School Science and Mathematics,* Fall, 1976.

The generally accepted scope and sequence of mathematics for young children involves the operations of addition, subtraction, multiplication, and division. Each operation, beginning with addition, is "taught" to children and after a few months of "practicing" the operation, they are introduced to another operation. This "taking turns" approach with each operation tends to emphasize an artificial separateness in the minds of the children, instead of stressing the interrelatedness of these "basic" operations. Textbooks, as much as anything else, tend to perpetuate the "taking turns" approach as well as another "mind narrowing" process. Most textbooks and workbooks also tend to emphasize *form* over *process*. For example, $3 + \square = 7$ suggests among other things, that addition is a left to right process. Unless many forms of the same process are introduced simultaneously, children tend to develop an unnecessarily narrow interpretation of the process of addition. The forms

$$3 + \square = 7,\ \square + 3 = 7,\ 7 = 3 + \square,\ 7 = \square + 3,\ \frac{\begin{array}{r}3\\+\square\end{array}}{7},\ \frac{\begin{array}{r}\square\\+3\end{array}}{7},\ \text{etc.,}$$

are all equivalent mathematical statements.

The most important message in this paper is not about how to reduce the emphasis on form, however. It is a proposal that will meet (and already has met, in discussions with my math education colleagues) with considerable resistance.

PROPOSAL: *Various forms of the four operations of addition, subtraction, multiplication, and division should be totally interrelated in a concrete way from the very beginning, when a child is introduced to "mathematics."* The proposal is based on two things: (1) a personal conception of the nature of mathematics, and (2) certain aspects of the developmental psychology of Jean Piaget and his Genevan colleagues. We will look briefly at these underlying assumptions.

the nature of mathematics

Stated briefly and succinctly, "mathematics is the process of searching for patterns." The key words in this definition are *process, searching,* and *patterns.* "Process" suggests an ongoing activity in which the answer or product is deemed secondary to the actual process. "Searching" suggests an emphasis on self-directed behavior as opposed to following already established procedures. And the word "patterns" suggests links between common elements in a system that unify the elements into a more highly interrelated "whole."

Using this definition of mathematics, curriculum developers and teachers would emphasize certain things (about content and teaching methods) and de-emphasize others.

Piaget's work

The nature of intellectual development in children and theories about such development should provide direction to educators who are interested in helping children develop an understanding of mathematics. Piaget's work has provided the lion's share of our present knowledge about how children gradually develop logical abilities.

One aspect of Piaget's theory of cognitive development, one that is stressed over and over in his writings, is the *interrelatedness* of his hypothesized mental structures. *Development proceeds as an interrelated whole.* Classification, for instance, is inextricably related to seriation. Although we may speak of two separate processes, the human organism, in fact, operates as a total, interrelated whole.

The concept of number and the various logical operations that we associate with mathematics develop, at various levels, in an interrelated way. The operation of addition, is seen, both developmentally and mathematically, as the inverse operation of subtraction, and vice-versa. Multiplication and division have a similar relationship. In fact, all four "operations" are totally interrelated developmentally and mathematically. To communicate anything else is developmentally and mathematically incorrect.

Piaget has identified the processes or operations of classification and seriation as the underlying logical structures for beginning logic, and thus for beginning mathematical operations (see *The Early Growth of Logic* and *The Child's Conception of Number* for a more detailed account). Whereas different types of classification and seriation are achieved at various ages, the mental structures that form the basis for these operations are developing as an interrelated whole. The concept of "numberness" is gradually evolving as the child interacts with the environment and has opportunities for continued growth. Piaget's own ideas regarding the interrelatedness of these basic logical structures are summarized below as a somewhat lengthy but remarkably clear (for him) quotation:

> It is wrong to think of additive structures as being established first and then generalized, so producing multiplicative structures of two or more dimensions.

In point of fact, at every level we find children using some form of classification, however rudimentary, and each of these can be applied to several criteria simultaneously as well as to one alone. In the first case the structure is multiplicative, in the second it is additive, but there is no essential opposition between the two.

That these two structures develop through parallel stages, and in close mutual dependence, shows that they constitute a single operational organization, in spite of differences in graphic power and in complexity. We shall also find a genetic relationship between the system of ordering and that of classification. In our opinion, this sort of genetic connection constitutes one of the weightiest arguments in favour of an operational conception of intelligence.[1]

Taken at the same level of sophistication, subtraction is no more or less difficult than addition. This statement (experimentally, a hypothesis) is derived from developmental theory rather than an analysis of the "discipline" of mathematics. Before the reader begins to develop mental arguments against the statement, let him answer this question: "Do you know of any carefully controlled studies in which children were "taught" the process of subtraction well before addition was introduced?" I think not. From the standpoint of set theory at an intuitive level, it is no more difficult to remove elements from a set than to add them. One process is merely the inverse of the other. The concept of "sixness" can be derived equally well from actions which parallel "addition" or "subtraction." There is nothing inherently easier about $4 + 2 = 6$ when compared to $7 - 1 = 6$. It only *seems* to be easier because addition nearly always precedes subtraction in our elementary classrooms. We expect it to be easier and communicate these expectations to children in a myriad of ways. Based on both intuitive experiences in our environment and on logic, understanding the concept of "sixness" is equally easy (or difficult) through the processes of addition or subtraction.

Division of a set of six marbles into two sets of three is likewise no more or less inherently difficult than adding $4 + 2$ or subtracting 1 from 7. The basic intuitive concept of "dividing" a group of things into two or more groups can be approached at a difficulty level equal to that of addition or subtraction. Before a child is "taught" $3 + 3 = 6$, that child has no developmental basis for understanding addition more easily than subtraction, at the same difficulty level. For example, 3 marbles moved together with another 3 marbles is no more or less difficult than dividing a set of 6 marbles into two groups of 3 marbles. As a mathematical procedure, division appears to be more difficult than addition or subtraction because *we have said it is.* If division is introduced two or three years after addition, then the natural assumption by all involved (including children) is that it must be considerably more difficult. The same arguments can be applied to multiplication, but for the sake of space, it is assumed that the reader can extrapolate the procedures to include this process.

The upshot of all of this is that our traditional sequence of mathematics for young children has no real basis, in fact, either developmentally or mathematically. The Piagetian buff may be yelling "multiplicative classification" and "multiple seriation" at this point in disagreement over the comparative difficulty of, say, addition and multiplication. It must be remembered, however, that what has been argued so far has been the basic interrelatednes of addition, subtraction, multiplication, and division at an intuitive, concrete level. The child's actions with objects (marbles, etc.) can equally well reflect the basic properties of any of the operations. A specific form of subtraction *can* be more difficult than addition but the reverse can also be true.

At the same difficulty level, for the concept of "sixness" where only two groups of objects are involved, addition and subtraction as active operations are interchangeable. Multiplication and division, of course, can also be used in the analogy. When the symbols (2, 3, 4, +, −, etc.) are derived from the actions of the child, it is the developmental level of the child, not the traditional sequence of mathematics curricula, that determines what makes sense.

The problem of teaching "form" rather than process, which was brought up early in this paper, is an outcome of emphasizing the wrong thing in mathematics for children. When addition is taught in isolation from the other processes of subtraction, multiplication, and division, the curriculum is at odds with the interrelated nature of cognitive development. And because they are taught separately, a kind of separateness is communicated to the child.

summary and recommendations

The traditional sequencing of the operations of addition, subtraction, multiplication, and division in our elementary schools is not founded on solid evidence. Further, this sequence tends to communicate an artificial separateness and, therefore, reduces the natural interrelatedness (developmentally and mathematically) among these logical processes. The concept of, say "eightness," is greatly extended when experienced from a variety of processes and forms.

"Mathematics as a process of searching for patterns" takes on far more meaning when approached from a broader, more interrelated conceptualization of patterns and processes. The concept of "eightness," when experienced as addition, subtraction, multiplication, and division from the beginning, promotes a greater understanding of their interrelatedness on the part of the child. Incorporating the following recommendations into the regular mathematics curriculum would tend to bring it more in line with our current understanding of the early growth of logic in the child.

1. In using manipulative materials with children, emphasize the *actions* associated with addition, subtraction, multiplication, and division of sets of objects. The symbolic notation (1, 2, 3, +, −, etc.) should never precede an operational understanding of these processes, as evidenced by the children's actions and responses to questions.
2. Searching for patterns should be central in all activity. For example, how many different ways can be found to "produce" six marbles? For addition: $5 + 1$, $4 + 2$, $3 + 3$. For subtraction: $7 - 1$, $8 - 2$, $9 - 3$, etc. For division: $12 \div 2$, and so on. For multiplication: $2(3)$, $1(6)$, $3(2)$, and so on. All operations, of course, are done at a concrete rather than a symbolic level until children give indications that they are ready for such "shortcuts." Many types of measuring activities should also be done to generalize the concept of "sixness." Taking five steps and then one more to arrive at six, backing up one from seven, three jumps of two steps each, and so on, all get at the concept of "sixness" through concrete representations of various operations. The interrelatedness of these operations will become more and more apparent as the children experience more ways of arriving at the same place. Form will be de-emphasized in favor of function or process. Many right answers will be offered because of the many ways of solving the "problem."
3. As children become more able to symbolize their actions, care should be taken to interrelate the processes of addition, subtraction, multiplication, and division as different procedures for solving similar problems. One process is often more appropriate than another for a given problem, but the important thing is to realize the interrelatedness of all possible solutions. The level of difficulty progresses with all four processes going forward. The concept of "twelveness" is intuitively more difficult than "sixness" and, mathematically, represents an extension of the number system. There are more opportunities for pattern making and likewise, more opportunities for using the four "fundamental" logical operations of addition, subtraction, multiplication, and division.
4. Research efforts should be undertaken to test the proposals in this paper. What effects occur when the traditional sequencing in mathematics for elementary schoolchildren is discontinued in favor of the interrelated approach outlined in this paper?

Some possible "designs" for such studies would be these:

a. Begin with subtraction for group *A* while group *B* receives an identical treatment, except these children begin with addition.
b. Use the operations of addition and subtraction simultaneously as first experiences to mathematics (group *A*) and compare the effects with a control group (*B*). Obviously, all other factors, except this sequencing factor, should be the same for both groups.
c. Interrelate the operations of addition, subtraction, multiplication, and division as first experiences to mathematics (group *A*) and compare effects with a "begin with addition only" (group B).

It is important to emphasize the necessity of controlling for such factors as teaching methods, use of manipulative materials, affective environment created by the teacher, student characteristics, and any other variables which are known to be influential in affecting certain outcomes in children.

Any logical progression of curriculum development and implementation would always be preceded by recommendation four—research. That seldom seems to be the case in actual practice, however. If the proposals in this paper were verified and implemented, substantial changes would be seen, at least in the publishing industry. It is usually very difficult to remember how certain "traditional" educational practices came into being. The question of sequencing the operations of addition, subtraction, multiplication, and division over a period of a few years is no exception. Our current knowledge of the development of the beginnings of logic in the child as well as a logical analysis of the operations themselves form the basis of an argument that is sufficiently strong to question the present way of doing things. It is hoped that some of the research-oriented readers of this journal will agree and begin to research the issues carefully and objectively.

references

INHELDER, B., AND J. PIAGET. *The Early Growth of Logic in the Child.* New York: W. W. Norton & Co., 1969, p. 195.

appendix c

The Myth of Stages In Cognitive Development*

* This paper by Ronald G. Good is here published for the first time.

the myth of "stages" in cognitive development

There are no real stages of development. There are merely categories made up by people for the purpose of communicating ideas in a less complicated manner. It is this basic assumption that I will try to explain and justify in this paper.

Jean Piaget has come to be associated with the field of developmental psychology in much the same way that Albert Einstein was associated with relativity theory in physics. Both associations are certainly justified. Because of the nature of the field of study and partly because of Piaget's tendency to write in a somewhat mysterious style (at least to the beginner), there are different interpretations as to the meaning of his work. The concept of "stage" has, in my opinion, been widely misinterpreted. The ever-increasing supply of "introductions" to Piaget's research and theory tends to perpetuate the general notion of stages of development. Oversimplification often leads to miscommunication and such is the case with the concept of "stages" in cognitive development theory.

Four general stages of development have come to be associated with Piaget's ideas: (1) sensory-motor (zero to two years), (2) preoperational (two to seven years), (3) concrete operational (seven to eleven years), and (4) formal operational (eleven to fifteen years). Certain cognitive characteristics are associated with each stage, and various experiments have been performed to show the kinds of responses children give during the interview. Most of the more recent work has been done with children beyond three and four years of age, so most of the examples used here will deal with that range.

The conservation studies are a good place to begin. Countless studies have been reported which show that children "conserve" various properties (number, mass, length, area, weight, volume, etc.) in a certain sequence and at generally specified age ranges. The "conserver" is able to use some form of logic to overcome perceptual "miscues" in an interview situation using various objects. More and more evidence has shown that wide variations exist in the responses of children to conservation interviews depending on a variety of factors (Izgiris; Pratoomraj and Johnston; Gruen; Minichiello and Goodnow; Nummedal and Murray; Good). These factors include (1) nature of the materials used during the interview, (2) nature of the questions asked by the interviewer, and (3) criteria used to define *conserver*. So what does it mean to say that a child is a conserver of something? It appears to mean that conserver has as many different meanings as there are tasks devised to measure the subjects' responses. The level of difficulty of an interview task can vary widely, as evidenced by the wide variations in results with similar groups of children. Even adults who have been labeled as conservers of something seem to revert to nonconservers under certain conditions (Lawson and Renner, 1974). If a person is a conserver of area (as defined by a certain interview task) and a nonconserver of weight, should we refer to the person as preoperational or concrete operational? What of the adult who is a nonconserver of weight or one who "reverts" to nonconserving behavior under certain conditions? Is it possible that the word conserver has meaning only to the extent that each experimental condition is carefully described? If so, how can we use the word conserver to define a "stage" of thought, say concrete operations?

Let us consider the stage of formal operational thought for a moment. Various tasks have been used to gather data on adolescents and adults, although the studies have not been as numerous as those with younger subjects. Nevertheless, it is becoming just as clear that wide variations occur in subjects' responses to various interview tasks at the so-called formal level. Interview tasks devised by Piaget and his coworkers (McKinnon and Renner, 1971) have been used with widely varying results in more recent investigations (Minichiello and Goodnow, 1969; Nummedal and Murray, 1969; Pratoomraj and Johnston, 1966). So even with the same tasks (or very nearly the same), the percentages of subjects categorized as "formal operational" vary widely. It seems logical to assume that devising tasks with greater "apparent" difficulty would result in fewer subjects being categorized as formal operational. In other words, we could have as many different levels of formal operational thought as we have tasks of varying difficulty.

The "stage" of concrete operations or the "stage" of formal operations begins to lose any real, consistent meaning when interpreted as suggested previously. Development is continuous and stages are devised so as to simplify communication. Stages do not fit the data. To speak of formal operational thought (or anything else) without specifying the conditions of data collection and their interpretation is about as vague as referring to an I.Q. score. This is probably not earthshaking news to the serious student of Piaget's work who has spent a few years reading and interacting with children and adults. For the other 90-plus per cent of the population, however, it is very likely that "stage" development is seen as a reality in intellectual growth. Those researchers and writers who are currently "interpreting Piaget's work for the masses" or reporting their own results can do a great deal to help the student of cognitive development theory gain a more accurate understanding of the limitations and drawbacks of interpreting the stages concept literally. To emphasize the importance of this issue, let me describe a recent incident.

At a seminar in my department, a visiting faculty member, who had recently received his Ph.D. in science education, described his dissertation research. It dealt with the use of Piaget's work to assess certain aspects of a middle school science curriculum project. A set of pretests for "formal operational" thought, which have been described by Piaget, were used and supposedly "matched" with certain content from the curriculum project. The matched tasks were used as posttests and it was found that students did less well after the treatment of the curriculum project than before. The interpretation was that the experiences associated with the curriculum somehow caused students to regress in ability to use formal operational thought. At that point another interpretation of the lower posttest scores was offered by me, namely, that the posttests were more difficult than the pretests. This interpretation was rejected immediately because "the tests *had* to be of equal difficulty since the theory (Piaget's) predicted that they should be."

I believe it is clear to the reader that a narrow interpretation of the "stage" of formal thought and tests that measure this stage were probably responsible for a less than objective interpretation of the data. Different tests will very likely measure different things no matter how similar the theory predicts they should be. The reason the previous example was used in this paper is to illustrate the misleading effect of assuming a singular definition of a stage of cognitive development. This misleading effect is not limited to the casual reader of Piaget's work, as we have just seen. Jean Piaget and his colleagues have provided us with an extremely valuable set of ideas about the intellectual development and functioning of human beings. These ideas should be closely scrutinized and reconsidered continuously, as should all theories. We should guard against the tendency to accept the "givens" such as "stages" of cognitive development because it limits our ability to interpret new data and reinterpret old data in unique ways. Development is clearly a continuous process and the choice of the number of stages to be used in describing the process is arbitrary. It logically follows that there is no single definition of concrete operations or formal operations. There are as many definitions as there are means to define.

references

BART, W. "The Factor Structure of Formal Operations," *British Journal of Educational Psychology,* **41,** 70–77, 1971.

GOOD, R., ET AL. "Cognitive Development Characteristics of Children in Grades 1–5: Evidence for Questioning the Generalizability of Conservation Data," paper presented to the 46th annual meeting of the National Association for Research in Science Teaching, Detroit, 1973.

GRUEN, G. "Note on Conservation: Methodological and Definitional Considerations," *Child Development,* **37**, 977–983, 1966.

HALL, V., AND R. KINSLEY. "Conservation and Equilibration Theory," *Journal of Genetic Psychology,* **113**, 195–213, 1968.

INHELDER, B., AND J. PIAGET. *The Growth of Logical Thinking: From Childhood to Adolescence,* New York: Basic Books, 1958.

IZGIRIS, I. "Situational Generality of Conservation," *Child Development,* **35**, 831–841, 1964.

LAWSON, A., AND J. RENNER. "A Quantitative Analysis of Responses to Piagetian Tasks and Its Implications for Curriculum," *Science Education,* **58**, 545–559, 1974.

MCKINNON, J., AND J. RENNER. "Are Colleges Concerned with Intellectual Development?" *American Journal of Physics,* **39**, 1047–1052, 1971.

MINICHIELLO, M., AND J. GOODNOW. "Effect of an Action Cue on Conservation Amount," *Psychonomic Science,* **16**, 200–201, 1969.

NUMMEDAL, S., AND F. MURRAY. "Semantic Factors in Conservation of Weight," *Psychonomic Science,* **16**, 323–324, 1969.

PRATOOMRAJ, S., AND R. JOHNSTON. "Kinds of Questions and Types of Conservation Tasks as Related to Children's Conservation Responses," *Child Development* **37**, 343–353, 1966.

appendix d

A Self-test

It is suggested that you attempt to answer these questions before beginning to study the material in this book and then again after completing the book.

1. How did Piaget's professional training as a biologist influence his later work in the investigation of the development of thought?
2. What do the following types or categories of children's responses to interview questions suggest about their thinking?
    a. Answer at random.
    b. Romancing.
    c. Suggested conviction.
    d. Liberated conviction.
    e. Spontaneous conviction.
3. How do interviews with manipulative aids compare with interviews not using such aids? Give specific examples.
4. What are the essential characteristics of a "genuine" class and how does classification ability develop in children?
5. How are classification and seriation related to the development of an understanding of the number concept?
6. How do children develop spatial concepts from the ages of three or four years to ten or eleven years?
7. What are the relationships among topological, projective, and Euclidean space?
8. If a child lacks a coordinated spatial perspective, what would be some possible limitations to his/her learning potential?
9. What does it mean to say that a person is a "nonconserver" of something?
10. How would one go about finding if a person is a conserver of length? Area? Number? Volume? Weight?
11. How are the concepts of time and motion related, in terms of intellectual development?
12. How do the cognitive characteristics of a five- or six-year-old compare with those of a nine- or ten-year-old? Use specific examples.
13. Why are such concepts as probability, correlation, and proportion considered to be "formal" or abstract?
14. Why are the processes of identifying and controlling variables so important in the field of science?

15. How could one determine a person's ability to systematically identify and control variables in an experimental situation? Give specific examples.
16. What is the relationship between Piaget's INRC group and propositional logic?
17. How does concrete thought (concrete operations) compare with abstract thought (formal operations)?
18. How is Piaget's *theory* of cognitive development related to his *research* into children's thought?
19. How does "science as content" compare with "science with process"?
20. How do the intellectual characteristics of children in the first grade compare with the conceptual requirements of such science concepts as weather causes and our solar system?
21. How do the intellectual characteristics of children in third grade compare with the conceptual requirements of such concepts as energy sources, the nature of sound, and chemical changes?
22. How do "concrete" science concepts compare with "formal" science concepts, in terms of cognitive development? Give specific examples.
23. What are some ways to *objectively* analyze science text material?
24. What are some specific *processes* that might be used to define science?
25. What science processes might be consistent with the intellectual characteristics of children in first grade? Inconsistent?
26. What science processes might be consistent with the intellectual characteristics of children in fifth grade? Inconsistent?
27. What definition of science would be consistent with what is known about the cognitive (intellectual) characteristics of children?
28. What are a few important implications for helping children learn science that grow out of cognitive development theory and research?
29. What are a few distinctly different instructional roles that could be assumed by the elementary school science teacher?
30. Which teacher roles (instructional strategies) are compatible with what is known about the intellectual development of children? Incompatible?
31. How would one go about determining whether a particular science program is compatible with the conditions required to help children learn science?

32. What type of physical setting would facilitate the process of helping children learn science?
33. How might a child in first grade try to find out about the growth of plants? A fifth-grade child?
34. How might a child in the third grade try to find out about the nature of insects?
35. What types of outdoor science activities would be consistent with the conditions necessary to help young children learn science? Older children?
36. What are the conditions required (including teacher role) to maximize the possibilities for helping children learn science?
37. Can you help children learn science?
    a. If yes: What evidence do you have to support your contention?
    b. If no: What can you do to learn how?

index